koe

koe

An Aotearoa ecopoetry anthology

Edited by

**JANET NEWMAN &
ROBERT SULLIVAN**

OTAGO UNIVERSITY PRESS
Te Whare Tā o Ōtākou Whakaihu Waka

Contents

- 9 FOREWORD
- 11 INTRODUCTION
- 29 THE EARLY YEARS: *Ecopoetry by poets born in the nineteenth century or before*
- 38 Hutia te rito o te pū harakeke, *Patricia Grace and Waiariki Grace (transl.)*
- 39 Titiro kau ana ko ngā pari pōhatu, *Patricia Grace and Waiariki Grace (transl.)*
- 40 Ruia, Ruia, Tahia, Tahia, *Tūmatahina*
- 41 Ko te Moana, Ehara Rawa i te Wai Kau / The Sea is Not Any Water, *Provenance unclear*
- 42 He Waiata Whakaaraara Pā (A Sentinel's Song), *Pua-Roro*
- 44 He Tangi mo te Matenga o Ngā Kai (Lament for a Failed Crop), *Horomona Hapai*
- 46 Kaore Hoki Taku Manukanuka, *Te Kooti Te Arikirangi Te Tūruki and Pāora Te Pōtangaroa*
- 55 To the Makomako, or Bell-bird (Now rapidly dying out of our land), *Alexander Bathgate*
- 57 The Forty-Mile Bush, *Anne Glenny Wilson*
- 58 An Old Chum on New Zealand Scenery, *George Phipps Williams and William Pember Reeves*
- 61 The Passing of the Forest, *William Pember Reeves*
- 64 A Bush Section, *Blanche Baughan*
- 72 In London, *Dora Wilcox*
- 75 Pause, *Ursula Bethell*
- 76 Drought, *Francis Hutchinson*
- 77 Ki Kō, Ki Kō, *Te Māreikura Hori Enoka*
- 79 Dead Timber, *Alan E. Mulgan*

- 81 THE MIDDLE YEARS: *Twentieth-century ecopoetry*
- 91 Last Song, *Fleur Adcock*
- 92 Waipatiki Beach, *James K. Baxter*
- 94 Beginnings, *Peter Bland*

95 Conversation with a Ghost 1974–1985, *Arapera Hineira Kaa Blank*
97 The Land and the People (III), *Charles Brasch*
98 The Return, *Alistair Te Ariki Campbell*
99 Our Tūpuna Remain, *Jacq Carter*
100 House and Land, *Allen Curnow*
102 Pioneer Woman With Ferrets, *Ruth Dallas*
103 The Old Place, *Harry Dansey*
104 Atom Bomb Test, Moruroa Atoll, 6 September 1995, *Lauris Edmond*
105 Otago Landscape, *Tricia Glensor*
106 Lake Manapouri, *Denis Glover*
108 Waikato te Awa / Waikato is the River, *Rangi Tanira Harrison*
112 Hope, *Dinah Hawken*
114 Blaketown Beach, *Noleen Hood*
115 Three Pines by the Hohonu: A celebration, *Peter Hooper*
117 Te Rapa, Te Tuhi, me Te Uira (or Playing with Fire), *Keri Hulme*
120 A Bottle Creek Blues, *Sam Hunt*
122 Beginning Again, *Donald McDonald*
124 The Mess We Made at Port Chalmers, *Cilla McQueen*
125 Aramoana, *Hirini Melbourne*
126 Papakainga, *Trixie Te Arama Menzies*
127 The Road, *Barry Mitcalfe*
128 Te Kōkota o Pārengarenga / The White Sands of Pārengarenga, *Saana Murray*
130 Eldorado Poem, *John Newton*
132 Ornamental Gorse, *Chris Orsman*
133 Waikato Lament, *Evelyn Patuawa-Nathan*
134 Tāmaki-makau-rau / Tamaki of a Hundred Lovers, *Merimeri Penfold*
136 Song from Kapiti, *Vernice Wineera Pere*
138 A Cloak and Taiaha Journey, *Roma Potiki*
140 Memo for Horace, *Harry Ricketts*
141 The Bomb is Made, *Keith Sinclair*
142 As the Godwits Fly, *J.C. Sturm*
144 Feelings and Memories of a Kuia, *Apirana Taylor*
147 Mururoa / Moruroa, *Ngahuia Te Awekotuku*
149 Ore: An ecological poem, *Denys Trussell*
151 Van Morrison in Central Otago, *Brian Turner*
153 No Ordinary Sun, *Hone Tuwhare*
154 Papa-tu-a-Nuku (Earth Mother), *Hone Tuwhare*

155 Haka: He huruhuru toroa / Haka: The feathered albatross, *Muru Walters*
158 *from* Pathway to the Sea, *Ian Wedde*
161 In the Midnight Ocean of his Sleep, *Albert Wendt*

163 NOW: *Twenty-first-century ecopoetry*
172 Seal Mourning, *Jenny Argante*
174 Marginalia, *Bridget Auchmuty*
175 Fucking St Barbara (i), *Tusiata Avia*
176 Huia, 1950s, *Hinemoana Baker*
177 Trout / *Oncorhynchus mykiss* / *Salmo trutta*, *Airini Beautrais*
180 Feature Battle, *Brenda Burke*
181 If I Am the River and the River is Me, *Jacq Carter*
183 Ghost Stoat, *Jonathan Cweorth*
185 Speaking to the Otter, *Lynn Davidson*
186 A Report on the Ocean, *David Eggleton*
188 Losing our Mana, *Rangi Faith*
189 Eel, *Fiona Farrell*
193 Story Lines, *Sue Fitchett*
196 Candle, *Alison Glenny*
197 Te Aro, *Jordan Hamel*
198 Losing Everything, *Dinah Hawken*
199 The Land without Teeth, *Rebecca Hawkes*
202 The Anthropocene *circa 2016*, *Helen Heath*
205 Old Bones, *John Howell*
206 Recipe for a Unitary State, *Gail Ingram*
207 Moruroa: The name of the place, *Kevin Ireland*
209 Huia, *Anna Jackson*
210 2050, *Ash Davida Jane*
212 All That Summer, *Tim Jones*
214 Flood Monologue, *Anne Kennedy*
220 Phosphate from Western Sahara, *Erik Kennedy*
221 I. Pūhā, *Megan Kitching*
222 *from* Death of a Landscape, *An Aside:* Advice to the Rehabilitator, *Leicester Kyle*
224 Birdspeak, *Arihia Latham*
226 Oh Dirty River, *Helen Lehndorf*
228 Korari / Harakeke, *Toi Te Rito Maihi*

229 Huia, *Bill Manhire*
230 Girl from Tuvalu, *Selina Tusitala Marsh*
232 Ends, *Carolyn McCurdie*
235 How They Came to Privatise the Night, *Maria McMillan*
237 Frogs, *Cilla McQueen*
238 There's Real Mānuka Honey in Heaven, *Van Mei*
239 Matariki: A call to kāinga, *Karlo Mila*
243 Dear ET, *Harvey Molloy*
244 Birdlife in a Broken Century, *Elizabeth Morton*
246 Huia, *Emma Neale*
248 So You Don't Belong, Pohoot, *Naomi O'Connor*
249 Piha, *Kiri Piahana-Wong*
251 Praise the Warming World (Try to), *Robyn Maree Pickens*
252 Whale Fall, *Nina Mingya Powles*
254 Topside, Nauru, *Vaughan Rapatahana*
256 The Old Breed, *Richard Reeve*
257 The Rena, *Te Kahu Rolleston*
259 Erasure, *Sam Sampson*
260 Dad's Piece of Sky, *Tim Saunders*
261 Port Hills, Canterbury, *Elizabeth Smither*
262 River Songs – Waimāpihi, *Ruby Solly*
264 Choosing, *Jillian Sullivan*
265 my whenua, *Apirana Taylor*
266 Mataura Paper Mill, *Anthonie Tonnon*
268 Lament for the Taieri River, *Brian Turner*
270 Pupurangi (Kauri Snail shell), *Hone Tuwhare*
272 Manawatū, *Tim Upperton*
274 Whai, *Briar Wood*
275 A Behoovement, *Sue Wootton*

277 NOTES
280 ABOUT THE POETS
292 SOURCES
297 FURTHER READING
298 ACKNOWLEDGEMENTS

Foreword

Koe refers to the cry of a bird. H.W. Williams' *Dictionary of the Māori Language* also defines 'koe' as a scream; a disturbing scream from the forest, or the shore, or the marshes, or the bush-lined gullies and gorges, or its absence from the paddocks, the eroded dead timber hillsides and mountains, the oxidation ponds and outfalls. Voices and isolated absences combine in this simple word.

Koe is also most commonly understood as the singular pronoun 'you'. The title of this anthology doesn't directly refer to 'you' – it isn't in a grammatical construction such as 'ko koe' (it is you) or 'tēnā koe' (greetings to you) – but the holistic poetics of the term suggests that it does. 'You' may be the speaker of the poem or the receiver of the poem. Neither of these, the speaker or the listener, you and I, have a bird's eye view of the rising and falling tides, earth and skies of these ecologically devastated islands. Yet we can hear the birds. We also have each other, just as the pronoun recognises the otherness of 'you.' Te reo Māori has other words for cry, such as tangi, and other ways of saying 'you', such as kōrua or koutou. Yet poetry is a personal, intimate artform. It does have a public voice, but its most emotionally affecting moments are private, 'small holes in the silence' as Hone Tuwhare puts it, or ripples that are 'wind huffed'. So the title of our anthology is *Koe*.

Poetry can be a simple, lyrical and emotionally centred recording technology. The images conjured up by this collection of poems and voices are those we believe reveal the ecological cares and concerns of nearly a millennia of poetic statements in te reo Māori and, more recently, in New Zealand English. We refer you to Tā Apirana Ngata and Pei Te Hurinui Jones' enduring multi-volume collection of poetry, *Ngā Mōteatea*, for more compositions. There are many writers in te reo Māori, and we celebrate the continuum of Māori oral and written poetic forms from our hundreds of iwi and many more hapū throughout the country. The compositions that appear in this volume are gestures toward the huge flow of kōrero represented throughout the marae, the worlds of toi Māori, and the many discussions in print, such as the scholarly introductions to *Ngā Moteatea* or the expertise of Tā Tīmoti Kāretu in *Haka: Te Tohu o te Whenua Rangatira*, or Tā Tīmoti's more recent, and at times very beautiful, discussion in te reo Māori of waiata, haka, and of our finest composers, *Mātāmua ko te Kupu!*

The other term in this title, 'ecopoetry', has a more recent history within poetics and is unpacked in our introduction. I will say that at some point during settler-colonialism, or Westernisation, or neo-capitalist industrial farming, we (you and I) became distanced from Papatūānuku and stopped looking at Ranginui and bought the groceries at the supermarket instead of tending to kai with our own eyes and hands. Yet we hope that you find within these pages the connections to Papatūānuku, to Tangaroa and Ranginui that feed our hopeful poetics of belonging to each other and caring for these places, voices and homes.

Tēnā koe.
Robert Sullivan
April 2024

Introduction

Why this anthology?

'Can poetry save the earth?' asked American literary critic John Felstiner rather playfully in the title of his field guide to nature poems.[1] Of course, the answer is no. We know that poetry can't save the earth, or us, from the destructive effects of climate change. It couldn't protect Aotearoa New Zealand from the flooding brought by Cyclone Gabrielle in 2023. Or from the ruinous effects of forestry slash washed to the lowlands following the felling of trees on steep terrain. Or from the land erosion and slipping that has occurred since nineteenth-century European colonials began felling indigenous forests. It can't stop the loss of roads, bridges, homes, lives. Yet, as Christchurch poet Gail Ingram writes, '[poetry] moves us … by bearing witness to pain, joy, all that our community is … it gets us back to living again.'[2] Aotearoa New Zealand ecopoetry 'gets us back to living again' by paying attention to nature and our place in it.

Koe: An Aotearoa ecopoetry anthology charts the genesis, development and heritage of this country's ecopoetry from an undefinable point in Māori settlement until today. It takes a developmental approach by choosing and contextualising samples of ecopoems from different periods, including translations into English of poetic compositions in te reo Māori.*

The collection is organised into three chronological sections: the early years, the middle years, and the present. Each section provides samples of as many voices as possible from the great variety of ecopoems published in English (or in translation into English). Only one poem has been selected from each poet in each section, except for Hone Tuwhare, who has two poems in the middle years section. There are many other poets whose work is ecopoetical but unfortunately could not be included in this short anthology. We hope the poetry within these pages will inspire readers to seek them out.†

---

* With digital access to macrons, their use in Māori words is now commonplace. This anthology cites quotations and poems exactly as they appear and many did not include macrons at the time of publication.

† Poems' sources and a list of further reading at the end of the anthology direct readers to the other ecopoems these poets have published.

'The early years: Ecopoetry by poets born in the nineteenth century or before' is ordered according to the poets' birthdates. Where these are not known, the date of composition or publication has been used. This ordering gives a sense of the genesis and development of ecopoetry in traditional and early-twentieth-century Māori poetry translated from te reo Māori into English, and early European colonial poetry.

'The middle years: Twentieth-century ecopoetry' is ordered alphabetically according to the poets' surnames. This section shows the changes in attitudes of Pākehā poets, who began the century with a sense of estrangement from nature and ended it with a strong sense of connection. The work of Māori and Pacific poets writing in English enters this section from the middle of the century. It portrays cultural embodiment in nature alongside sorrow for ecological loss, and the associated loss of lifestyle and mana.

'Now: Twenty-first-century ecopoetry' is also ordered alphabetically by surname. Some poets in this section address the spectre of ecological apocalypse and some respond to this grim forecast by employing the poetic technique of erasure. Others express sorrow for nature's degradation alongside the hope of renewal. The ecologically damaging effects of colonisation reflected throughout Māori ecopoetry are acknowledged by some Pākehā and Pacific poets, as is the prospect of restitution.

Defining ecopoetry

Ecopoetry is a particular type of poetry that conveys different ways of conceiving relationships between nature and culture. The word 'ecopoetry' was first coined in America in the 1990s to describe ecological poetry that had its genesis in the 1960s environmental protest movements. Such poetry, also called 'ecopolemic', was a call to action for social and political change. It sought to inspire protection of the natural world from human degradation at a time when those holding political and industrial power seemed deaf to this notion.

With the formation of green political parties in Western Europe, America, Canada, Australia and Aotearoa New Zealand in the 1970s and 1980s, environmental activism moved from the counterculture to the mainstream.[3] There was a shift away from applying the term ecopoetry to ecopolemical poetry only. Some poets and critics argued that poetry of perception was more engaging to readers and, therefore, more likely to inspire awareness of the need to value and protect the natural world. Ecopoetry evolved into a more inclusive field, and the term began to be applied to poetry that spoke to the idea that nature is not only necessary for human physical survival but also for human mental wellbeing. This was a notion that had originated in the British Romantic tradition of the eighteenth century, and the field of ecopoetry broadened to include historical poems that connect nature with individual human consciousness by poets such as William Wordsworth (1770–1850) and John Clare (1793–1864).[4]

While there is not a single definition of ecopoetry that critics agree upon, all current definitions tend to have common denominators: ecopoetry investigates the relationship between people and the nonhuman world; it portrays nature with humility rather than a sense of superiority or domination; and it pays attention to, and often foregrounds, the nonhuman world. This last point can be understood as the reverse of anthropocentrism, which foregrounds the human world and assumes that it is only in relation to human beings that anything else has value.

Ecopoetry neither subjugates nor idealises nature and often exists within a context of awareness of human-caused loss or denigration of nature. It values nature not as a resource for human exploitation but as an interconnected part of human life at both physical and psychic levels.[5] This notion of interconnection and, more recently, entanglement between human and natural worlds is at the core of contemporary ecopoetry, which has evolved to portray nature as resilient despite human influence or as struggling to survive in a technological age.

The idea that people controlled the environment informed much polemical ecopoetry in the early twentieth century. From this time the term 'environmentalist' began to be applied to those who sought to preserve natural places from societal progress and consequent degradation. The term 'environment' manifested the idea of separation between people and nature and shifted attention from 'the real actions of human beings and focused it upon an abstraction that not only lacks agency and presence but whose very conjuring is a mystification.'[6] In more recent critical definitions of ecopoetry, the notion of 'environment' is often replaced by the concept of ecology, the study of the relationships between living organisms, including humans, and their physical environment, which recognises the interdependence between culture and nature. Still, a dualism exists in comprehending the two as separate entities.

One way to distinguish some contemporary ecopoetry from historical ecopoetry is its focus on global, rather than local, ecological concerns. These include climate change, loss of species diversity, pollution, urbanisation and the threat of planetary environmental disaster. But in many cases, the themes of historical ecopoems – nature's persistence, humanity's interdependence with nature, loss of connection between people and nature's cycles and rhythms, colonialism, deforestation, even species extinction – overlap with themes concerning contemporary ecopoetry. Because of these overlaps, ecopoetry provides a lens through which to look back at poems written when readers 'did not necessarily have an ecological perspective to think about them from'.[7] As a concept, ecopoetry provides a contemporary gauge to measure the ecological intent of historical poems.

If time offers one axis when considering how to define ecopoetry, geography provides another. The relatively new field of postcolonial ecocriticism challenges the foundational Eurocentric conception of dualism in ecopoetry: people on one hand and nature on the other, interconnected, even entangled, but separated by a distinction between human and nonhuman worlds.

Tangata whenua – people of the land (Māori) – and tangata Tiriti – people of the Treaty (non-Māori New Zealanders) – have characteristic ways of conceiving how people and nature interact, which makes the ecopoetry of Aotearoa New Zealand unique. Māori voices within Aotearoa New Zealand ecopoetry portray embodiment between people and nature and when culture is embodied in nature, cultural expression is a part of ecology.[8] Māori creation narratives place people in 'a genealogy of earth and sky'[9] that arises from whakapapa, in which people are kin with natural elements and creatures, and from spirituality, whereby natural elements are inhabited by ancestors. Culture is a part of nature in that human and nonhuman worlds are integrated into an inseparable oneness.

Aotearoa New Zealand ecopoetry

Aotearoa New Zealand ecopoetry derives from both the English tradition with a heritage in Romanticism and ecopolemic, and traditional Māori poetry stemming from diverse formal and informal compositions in te reo Māori. These two traditions, along with the voices of Pacific peoples and other non-European settlers to Aotearoa New Zealand, have produced a contemporary ecopoetry that increasingly includes te reo Māori and sometimes Pacific and Asian words and phrases in its English and, at times, bilingual lyric.

A comprehensive exploration of oral and written ecopoetry in te reo Māori is beyond the scope of this anthology, but it is important to acknowledge traditional Māori compositions as one of the intergenerational traditions from which the poetics and whakapapa of Māori poets writing in English can be traced. For readers wishing to learn more about these compositions, we suggest reading the multi-volume series *Ngā Mōteatea* edited by Apirana Ngata and Pei Te Hurinui Jones.[10]

Aotearoa New Zealand ecopoetry in English chronicles opposing and changing notions of ecological loss and belonging from the late nineteenth century until the present. The country's evolutionary history, geography and weather patterns, how people have interacted with the environment in these contexts, and the unexpected consequences of human action have produced very particular outcomes that have affected flora and fauna and, of course, the relationships between natural ecologies and culture.

The circumstances that have developed here can be related to the country's geological history. Environmentalist Tim Flannery describes Aotearoa New Zealand as 'a completely different experiment in evolution to the rest of the world' because 'birds occupied all of the major ecological niches occupied by mammals elsewhere'.[11] The pre-European country was a land of birds and ancient forest. Multiple and varied traditional Māori oral records pay close attention to nature and metaphorically connect human behaviours with those of indigenous species. Much later, ecopoetry by Pākehā poets focuses on colonial eradication of the forest and, sometimes, on the consequential loss of indigenous plants and birds.

Aotearoa New Zealand has a distinctive cultural history. The colonial hunger for land and assertion of economic and cultural dominance by Pākehā colonials over Māori was partly driven by the European perception of nature as external to the individual. This starkly contrasts the Māori worldview of culture embodied in nature. The colonial ethos to conquer nature and transform so-called wasteland into productive farmland, combined with capitalist notions of property ownership, was, and still is, in opposition with tūrangawaewae and kaitiakitanga. So, throughout its evolution, and especially since the regular publication of Māori perspectives from the 1950s onwards, New Zealand ecopoetry in English has explored different conceptions of how nature is used, occupied and imagined. The tensions between Pākehā and Māori views of nature and the human relationship with it are a defining characteristic of this country's ecopoetry.

Aotearoa New Zealand's colonial ecological transformation is distinguished by the speed and totality of change. It is further set apart by having occurred so recently that it is almost within living memory. Māori settlement around 800 years ago wrought ecological changes, but they were less extreme than the devastation that would ensue from European settlement. In the 100 years following the early 1800s, the imperial drive to incorporate Aotearoa New Zealand into the capitalist world economy led colonials to modify and, in the lowlands, annihilate indigenous ecosystems. This created what Eric Pawson and Tom Brooking call 'functionally incomplete systems dominated by introduced plant and animal species', which would 'trigger significant environmental problems within a generation'.[12] By 1920, 'the open country of New Zealand was a highly modified landscape. Indeed, the plant cover of most low country was almost unrecognisable from what it had been in 1840.'[13]

Deforestation, in concert with the country's weather patterns and a geography of central mountains and ranges, led to flooding and soil erosion. Overstocking and the wilful introduction and subsequent invasion of rabbits degraded much of the South Island hill country. Possums, stoats, exotic plants, and the unintentional release of weeds and pathogens devastated the remaining forests and bird life. Drainage of wetlands eradicated native species, creating what ecologist Geoff Park calls 'an imperial landscape ... of amnesia and erasure'.[14] He writes that in the land where most of us now live, 'one of humanity's most dramatic transformations of nature anywhere has removed [nonhuman] indigenous life almost entirely'.[15]

British colonialism nevertheless brought with it an appreciation of unspoilt, natural places. Areas perceived as untouched – despite Māori occupation – were considered exotic by European settlers and desirable as objects of adoration and retreat. Enabled by Crown confiscation of Māori land – and in some cases assisted by offers of land from Māori who, under duress, sought to protect taonga from sale to European colonials – an extensive conservation estate comprising some thirty per cent of the national land area was set aside as National Parks under the 1874 New Zealand Forests Act and the 1896 Urewera District Native Reserve Act. '[B]y 1900', write Pawson and Brooking, 'more and more people were valuing the indigenous as part of a growing identification with New Zealand as home.'[16] The strictly European view, which generally regarded those reserved areas as scenery providing a sense of solace and belonging, existed alongside but oblivious to the devastating effects of the loss of such land on iwi Māori and the ecological synergies on which whānau, hapū and iwi depended for the totality of their wellbeing.

The worldview of tangata whenua is described by Rev. Maori Marsden (Tai Tokerau): 'Man is the conscious mind of Mother Earth and plays a vital part in the regulation of her life support systems and man's duty is to enhance and sustain those systems.'[17] He describes people as 'an integral part both of the natural and spiritual order, for mauri [life force] animates all things'.[18] Mason Durie (Ngāti Kauwhata, Ngāti Raukawa, Rangitāne) also describes this worldview:

> Underlying the world views of indigenous peoples and at the heart of indigeneity is an 'ecological synergy spiral'. Basically about connecting relationships that are complementary and mutually reinforcing, the spiral moves from the small to the large, from individuals to groups, and from people, plants, fish and animals to the earth and sky.[19]

This understanding of human and nonhuman worlds as inseparable diverges from colonial Pākehā notions of nature as either a resource or a place of escape from modernity.

The nineteenth- and early twentieth-century export-driven expansion of farming not only transformed the countryside but considerably homogenised it. As Philip Steer says, 'It was on this violently modified and ever more industrial terrain that New Zealand literature took its increasingly distinctive shape.'[20] Undoubtedly, colonial ecological violence is in tension with the intentions of ecopoetry, and it was precisely in this environment that Aotearoa New Zealand ecopoetry in English had its genesis.

Early European settler poets followed the approach of British Romantic poets who focused on the beneficial effects of nature on the subjective self, that is, the sense of wellbeing individuals feel when they are in the presence of, or even remembering being in the presence of, nature. William Pember Reeves's 'The Passing of the Forest' (1898) recognises that along with the loss of trees and other plant and bird species through deforestation, 'the sense of noiseless sweet escape' is also lost. By comprehending nature as a source of solace from modernity, a place of escape from the hustle and bustle of commercial life, the poem tracks back to William Wordsworth's 'Lines Written a Few Miles Above Tintern Abbey' (1798), in which the memory of the view over the River Wye produces 'that blessed mood' where 'the weary weight / Of all this unintelligible world / Is lightened.' Another early Aotearoa New Zealand settler poet, Dora Wilcox, portrays the memory of this country's unspoilt bush from afar in her poem 'In London' (1926). The bush, she writes, is 'the valley of my Paradise', evoking a sense of pleasure in its recall not unlike 'Tintern Abbey' but with the addition of a sense of Christian fulfilment. And when Anne Glenny Wilson hears in 'this humming city' the 'music-breathing tree' of 'The Forty-Mile Bush' (1926), she contrasts, as Wordsworth did, the bustle of modernity with the consolation of nature's composure.

A Romantic approach is also evident in this country's twentieth-century ecopoetry. 'Three Pines by the Hohonu: A Celebration' (1972) by writer and conservationist Peter Hooper follows the tradition of American Romanticism with its echoes of Walt Whitman's 'Song of Myself' (1855). Whitman writes of nature: 'I am mad for it to be in contact with me.' Hooper's poem evokes a similar sense of bodily immersion by focusing on nature's beneficial effects on the individual. This is particularly evident in the final stanza in which 'I' appears eight times:

> I know I am clean and loved,
> I am companioned, I dive to the stones,
> I float, I wade and wallow.

> I am loosened, dissolved; delivered from time,
> I float in endless being, merge
> fulfilled in the body of the world.

Contemporary ecopoet Dinah Hawken also takes a Romantic approach in poems that evoke nature as a necessary component of human wellbeing in a technological age. 'Hope' (1995) portrays the growth habits of trees as a model that will benefit people in a time of grief:

> It is always bleak
> at the beginning
> but trees are calm
> about nothing
> which they believe
> will give rise to something
> flickering and swaying
> as they are: so lucid
> is their knowledge of green.

Throughout her oeuvre, in poems focused on natural elements, Hawken renders the composure of nature, 'the graceful things / that lift us up' ('Talking to a Tree Fern at Lake Rotoiti', 2001), as a model for people struggling to cope with the stresses of modernity.

Aotearoa New Zealand's most prolific Romantic ecopoet, Brian Turner, regularly evokes a psychic connection between his poet speaker and the rivers and ranges of backcountry Central Otago where he lives. He persistently portrays the region's natural elements as necessary to his speaker's emotional and physical wellbeing. 'Van Morrison in Central Otago' (1992), for instance, presents the mountains, rivers and sky as 'essential' to a sense of self:

> You have to be here, you
> have to feel the deep
> slow surge of the hills,
> the cloak of before, the wrench
> of beyond. You know
> what, you know
> not. And that's what
> makes it heart-stopping,
> articulate, hurtful
> like resuscitation.

Escape from modernity to a comforting nature is an enduring theme throughout Turner's work.

The Romantic poetry tradition is quite different in perspective from the Māori poetry tradition, which portrays communal integration with nature that is part of everyday life; a connection that has been articulated for hundreds of years in an oral tradition. Historian Arini Loader (Ngāti Raukawa, Ngāti Whakaue, Te Whānau-ā-Apanui) writes that in the 'nineteenth century Māori produced thousands of pages of written work … written almost exclusively in te reo Māori [that] enables unparalleled access to the first literature of Aotearoa-New Zealand'.[21] But until the 1950s, Māori perspectives were absent from this country's poetry in English due to an increasing English-language dominance in the early twentieth century and 'assimilatory ideology and government policy that severely undermined the health of the Māori language'.[22] Engagement with natural ecologies was powerfully expressed in other aesthetic domains, notably the carving and painting of meeting houses and oral narrative forms such as waiata and whakataukī.[23]

Traditional whakataukī translated into English reveal a prevailing theme of connection between people and nature through imagery and imagination. Patricia and Waiariki Grace write that the 'ancient proverbs, songs and chants' illustrate the Māori tradition that 'people are part of the universe, there being an interdependence among all life forms and all aspects of the physical and spiritual worlds' and that there is a 'need to live in harmony with nature rather than attempt to conquer and rule it'.[24] Some whakataukī speak directly to conservation:

> Hutia te rito o te pū harakeke   If you destroy the flax plant
> Kei whea te kōmako e kō?   From where will the bellbird sing?[25]

Others speak of ancestral connection with the land:

> Titiro kau ana ko ngā pari pōhatu   I look to the rock cliffs
> E whakaatu atu nei i ngā tīpuna.   And see the faces of my ancestors.[26]

These whakataukī portray the fundamental tenets of ecopoetry: a valuing of the natural world and awareness of the need to protect it . They also show an expansion of these tenets – attachment to place through whakapapa and kaitiakitanga.

A tradition of ecopoetry in te reo Māori was present in oral forms before and after European contact. It exists alongside poetry in English, which constitutes the Aotearoa New Zealand poetry canon or 'list of texts believed to be culturally central'.[27] Witi Ihimaera and D.S. Long note that: 'The Maori viewpoint has always been accessible to Maoris, that is, and to those who understand the Maori language; in this sense, it has been a "hidden" literature for many years'.[28] Translations into English of traditional waiata in te reo Māori include *Ngā Mōteatea: He maramara rere nō ngā waka maha / The Songs: scattered pieces from many canoe areas* (1928–88) collected by Sir Apirana Ngata with translations by Pei Te Hurinui, *Poetry of the Maori* (1961) translated by Barry Mitcalfe, *Maori Poetry: The Singing Word* (1974) collected and annotated by Barry Mitcalfe, *Maori Poetry: An introductory anthology* (1978) introduced and translated by Margaret Orbell, and *Ngā Tau Rere: An anthology of ancient Māori poetry* (2003) chosen by David Simmons and Merimeri Penfold.

Waiata, karakia, pātere, kaioraora, matakite, whakaaraara pā and haka are oral forms of poetry. 'The tradition of poetry in Māori society is far richer than that of prose', writes Te Kapunga Matemoana (Koro) Dewes (Ngāti Porou). 'Poetry is sung, intoned, recited and shouted.'[29] Indeed, Miriama Evans (Ngāti Mutunga, Ngāi Tahu) writes that Māori literature retains an oral rather than written form: 'Maori audiences continue to judge and appreciate contemporary Maori poetry primarily by its performance rather than its appearance in print.'[30] This anthology provides selected translations into English of early waiata, pātere, whakaaraara pā and whakataukī that reveal the ecopoetics of precolonial Māori. These poems portray a frame of reference or kaupapa that evokes an indigenous ethics of ecoconnectedness within a context of Māori protocols that guide relationships between people and all elements of the natural world. From this tradition, the heritage of Māori ecopoets writing in English, and published in English from the 1950s onwards, can be traced.

Although a triangulation between nature, colonial and Indigenous people is not specific to Aotearoa New Zealand, it does take a unique form here: the concepts of nature held by Māori and Māori terminology familiar to most New Zealanders contribute to this country's particular relationship with local ecologies. Anne Salmond writes: 'As Māori terms increasingly shift into New Zealand English, and vice versa, European and Māori ways of thinking alike are being transformed.'[31] Conceptual transformation is reflected in Acts granting legal personhood to the Urewera Ranges (2014), Whanganui River (2017) and Taranaki Maunga (2017). This status derives from Māori understanding of natural elements as kin with associated kaitiakitanga, reciprocal duties of guardianship between people and natural elements. According to environmental, Indigenous and human rights advocate Tina Ngata (Ngāti Porou), these Acts recognise a shift in colonial systems of conservation and care towards perspectives that are rooted in Māori ancestry and centred in rights of care other than ownership.[32] This inclusion of Māori concepts into law and advocacy frames how approaches to human relationships with ecologies are local and unique in Aotearoa New Zealand.

Māori relationships with local ecologies and the vital elements of nature – earth, sea and sky – are based on spiritual and genealogical connections leading to duties of guardianship rather than European notions of control and domination. These differences continue today, most visibly in Treaty claims over the use and perception of natural places. The 2014 deed of settlement between Whanganui iwi and the New Zealand government legally recognising the Whanganui River as a living entity – the first waterway in the world to gain this status – exemplifies how these cultural tensions overlap.[33] The Te Awa Tupua (Whanganui River Claims Settlement) Act 2017 gave the status of personhood to the Whanganui River. It included, amongst a list of intrinsic values to guide the governing body, the proverb:

E rere kau ai te Awa nui	My great river flows
Mai I te kahui Maunga ki Tangaroa	from the mountain to the sea
Ko au te Awa	for I am the river
Ko te Awa ko au	and the river is me

The proverb embraces what Gerrard Albert, chair of Ngā Tāngata Tiaki o Whanganui, sums up as 'rather than us being masters of the natural world, we are part of it'.[34] Its use in contemporary legislation reflects the significance and endurance of traditional te reo Māori narratives in the intergenerational transmission of knowledge and values.

Māori terms in New Zealand's Resource Management Act (1991) and Local Government Act (2002) requiring councils 'to take into account the relationship of Māori and their culture and traditions with their ancestral land, water, sites, wāhi tapu (ancestral sites) and other taonga', are also altering New Zealand law as the legal process is required to acknowledge 'the persistence and creativity of a distinctly Māori register of value'.[35] Such Acts, and 'a growing ease with terms like mana, tapu, utu, rangatiratanga and kaitiaki' highlight emerging recognition and 'a willingness by a non-Māori majority in New Zealand to recognise the value of Māori conceptions' and the extent of cultural-ecological differences.[36]

Until the 1950s, Māori and Pacific people's perspectives and voices were absent from New Zealand literature. Contributing factors were English language domination from the turn of the twentieth century, the disruption of traditional Māori communities because of contest between Māori and Pākehā land use values and practices, and the lack of recognition by Pākehā of mana whenua, the right of hapū to manage a particular area of land and to maintain relationships with ancestral land.

The publication of *Te Ao Hou*, a bilingual quarterly magazine by the Maori Affairs Department from 1952–76, provided a platform for Māori writers in English who were not recognised by the Pākehā poetry tradition or did not seek to publish their work in that environment. From the 1960s until his death in 1972, Pākehā poet James K. Baxter used his public image to raise awareness of the Māori world in the Pākehā establishment. His poetry connecting Māori and nature was openly political and an agent for social change.

But it was the Māori renaissance from 1972 that led to major transformations. Critic Melissa Kennedy describes this renaissance as:

> *the most significant literary movement since cultural nationalism in the 1930s and 1940s [which] asserted a separate nationalism within a bicultural nation, one with its own modes of expression, its own history, and its claim to represent a truly postcolonial Aotearoa-New Zealand.*[37]

Its governing tenets, including 'special status derived from priority in the land', have 'considerably influenced' New Zealand literature and literary criticism.[38] It led to a major cultural change in New Zealand, particularly from the time of the 1975 Hīkoi (Land March) and the founding of the Mana Motuhake Party in 1979. The first anthology of Māori writing, *Into the World of Light*, was published in 1982, and in 1991 Māori publishing house Huia was established. In descriptions of Māori connection to the land, novelists Witi Ihimaera and Patricia Grace made terms such as tūrangawaewae and tikanga nationally known and understood.

The first published collection of poetry by a Māori poet writing in English was Hone Tuwhare's *No Ordinary Sun* (1964). But it was Tuwhare's 'Papa-tu-a-nuku (Earth Mother)' (1978), written in support of the 1975 Hīkoi, that became widely recognised for its portrayal of the earth in human terms. Tuwhare's repeated use of the pronoun 'we' opposes the frequently used 'I' of Romantic poetry. His portrayal of a communal relationship between Māori and nature opposes the relationship between the subjective self and nature, on which prevailing definitions of ecopoetry are partly based. And in contrast with the Romantic notion of the natural world as a place of escape from everyday life, Tuwhare's poem evokes the land and people as kin:

> We are stroking, caressing the spine
> of the land.
>
> We are massaging the ricked
> back of the land

> with our sore but ever-loving feet:
> hell, she loves it!
>
> Squirming, the land wriggles
> in delight.
>
> We love her.

This poem epitomises the representation in Tuwhare's poetry of people as a part of nature and, crucially, of nature as a part of people. Reviewer John Huria wrote in 1993 that such representations position 'the self in a relationship of easy, close familiarity with nature, rendered in a way which may unsettle comfortable divisions between animate and inanimate'.[39] He argues that 'Papa-tu-a-nuku (Earth Mother)' 'relegates the notion of "personification" to the rapidly filling basket of Eurocentric redundancies. How can you personify a person?'[40] By portraying the earth as human, Tuwhare brings the worldview of Māori to Aotearoa New Zealand's poetry tradition in English. Jane McRae describes this as a shared genealogy with the physical world and its creatures, including beings from the ancestral world, all of which are kin.[41] The reactions of reviewers to Tuwhare's work – such as Bernard Gadd, who in 1984 described his use of 'imagery of a pantheistic-like animation of the natural world' – illustrated both the marginalisation of Māori poetry until the second half of the twentieth century and the differences between Māori and Pākehā worldviews that it exposed.[42]

Vernice Pere's 'Song from Kapiti', also published in 1978, compares the loss of the abundance of indigenous species with the loss of land, which pushed Māori to the margins. The remnants of indigenous life, 'sparse toe-toe' and a 'lone sea-bird' inhabit the same, cold cliff-face as the inhabitants of the Paekākāriki coast: 'I am that bird / frozen by the southern wind', Pere writes. The poem ends with the bleakness of alienation:

> I am the child of the Ngati-Toa,
> seeking my place
> in a mainland society.
> I am learning to sing
> the sad-sweet songs of a people's soul.
> I am the lone bird
> alive in a limbo of longing,
> enduring the winter world,
> surviving
> on the slim promise
> of a future summer.

Keri Hulme's novel *The Bone People* (1983) was a further pivot point, in which 'the unique Maori connection with the land is imagined as powerful enough to heal the modern-day ills brought by colonialisation and modernization'.[43] The 1985 Booker Prize winner located Māori spirituality in the land and in guardianship of nature.

In contemporary ecopoetry, poet and weaver Toi Te Rito Maihi's 'Korari / Harakeke' (2003) binds people with plants through merging the poet speaker with flax in 'a melding of the mauri of both … my fingers following multiplying images / within my mind …'. Pere's and Maihi's poems show that mātauranga Māori embodies culture in nature such that loss of natural elements and species is not only loss of ecologies but also loss of cultural expression.

Marginalisation of Māori literature continues into the twenty-first century. Witi Ihimaera and Tina Makereti note that Māori, Pacific peoples' and Aboriginal writing constitutes 'the disruptive act' in 'the worldwide literary landscape' and 'still the page is white, and still the marks we make upon it are radical acts of transgression, of forcing others to see us in all our complexity and wonder'.[44] Highlighting a contemporary lack of Indigenous perspectives in literature, they refer to the 2016 Te Hā Māori Writers Hui in Wellington where 'we talked about writing ourselves into existence'.[45]

Ihimaera and Makereti are the editors of *Black Marks on the White Page* (2017), a collection of 'Oceanic' stories and poetry that groups the work of Māori with writers from all over the Pacific, a writing community the editors describe as 'the same waka when it comes to literature'.[46] An Indigenous Oceanic worldview portrayed in contemporary Aotearoa New Zealand ecopoems by Tusiata Avia, Karlo Mila, Selena Tusitala Marsh and Robert Sullivan evokes cultural and ecological unity between Pacific peoples, including Māori, and the Pacific region centred on connection with Te Moana-nui-a-Kiwa, the Pacific Ocean.

Postcolonial ecocriticism, a relatively new field of study, recognises the importance of the local rather than the global and considers the relationships between literature and nature in postcolonial or settler countries, such as Aotearoa New Zealand. A postcolonial ecocritical framework contends that literature that has been sidelined by settler hegemony reveals comprehensions of nature and the human relationship with it that differ from prevailing narratives. It recognises that culturally specific ways of seeing the world disrupt and enlarge Western notions of ecological appreciation and encroachment.

This approach challenges the prevailing Eurocentric ideologies of ecopoetry, which are in tension with Indigenous understandings. Terms such as 'ecocentric' and 'nonhuman nature' that underlie present definitions of ecopoetry raise questions such as which ecologies and whose nonhuman nature? Rather than interdependence between people and nature, a phrase that currently underlies definitions of ecopoetry yet implies nature on the one hand and culture on the other, Aotearoa New Zealand ecopoetry portrays an overlapping of nature and culture. It foregrounds what DeLoughrey et al. describe as how 'the history of colonialism necessitates the imbrication of humans in nature' because 'postcolonial environmental representations often engage with the legacies of violent material, environmental, and cultural transformation'.[47] The notion of 'imbrication of humans in nature' suggests the way in which colonial culture transforms nature by introducing exotic ecologies, which over time become so familiar that they are perceived to be natural. In this way, people and nature overlap through physical change and perceptive regard. In this country's ecopoetry, colonially constructed nature, such as exotic plants and animals or waterways and landscapes that have been shaped by colonial transformation, are sources of Pākehā solace and contribute to a sense of Pākehā belonging. In contrast, elements of nature, including landforms, stars, wind, sea and indigenous plants and animals, are whakapapa, and their loss through colonial transformation is a source of Māori sorrow and contributes to a loss of a sense of Māori belonging.

*

This anthology shows the different ways Māori, Pacific peoples, Pākehā and non-European settlers conceive nature and the human relationship with it. It reveals changes in those conceptions, especially by Pākehā, over time. It portrays the ruinous effects of European colonisation on indigenous ecologies and Indigenous culture. It depicts attitudinal changes brought by the green movement of the 1960s when political activism generated ecopolemic in support of nature conceived not just as a resource but valued in its own right. It explores a Māori tradition of ecopoetry in te reo Māori alongside Pākehā perspectives in English in the nineteenth and early-twentieth century and shows how, from the middle of the twentieth century, the publication of the work of Māori and Pacific poets within the mainstream poetry canon conveyed, for the first time in English, culture as a part of nature. It reflects the evolution of the Māori cultural renaissance when voices previously sidelined or ignored by the country's English poetry tradition became integral to the largely Pākehā poetry establishment. And it reflects the significant, ongoing role of traditional te reo Māori narratives in contemporary responses to ecological protection.

Editor's statement

As a settler colonial, Pākehā, poet, literary scholar and non te reo Māori speaker in Aotearoa New Zealand, I am aware of the risks associated with translations of poems from te reo Māori to English and the risk of comparison. That is, the comparison of Māori poetry in response to the norms of English-language poetry risks aligning with approaches that 'reaffirm the centrality of the dominant institutional frameworks of Anglo-American and European literary studies'.[48] Indeed, within a centred Māori worldview, the concept of ecopoetry might seem redundant, given the fundamental integration of culture with nature. Despite this risk, the ability to challenge and expand ecopoetry's Eurocentric foundations seems important to the comprehension of Aotearoa New Zealand poetry and to the global field of ecopoetry. By challenging the Eurocentric foundations of ecopoetry, I hope this anthology exposes the limits of such dominant frameworks.

The idea for this anthology arose from research into the field of ecopoetry, which interested me because of its focus on the relationship between people and nature and its concern with protecting nature from human degradation. It became apparent that while ecopoetry is a relatively commonplace term in America and Europe, in Aotearoa New Zealand no critical studies that read this country's poetry from an ecopoetical perspective have, until now, been undertaken. This seems surprising given that a literary focus on the landscape has been well documented here.

As a descendant of settler colonial farmers, and a farmer myself, I am particularly interested in how ecopoetry negotiates perceptions of colonially constructed ecologies. Reading this country's poetry through an ecopoetical lens led quite quickly to the discovery of two traditions of ecopoetry: the te reo Māori tradition and the English tradition. It revealed different ways of perceiving indigenous and exotic ecologies, including the dire effects of species loss on Indigenous culture. In terms of belonging – and I feel a sense of belonging to my family farm – ecopoetry discloses how my sense of belonging through nature differs from a sense of belonging rooted in whakapapa and kaitiakitanga. It widens my comprehension of what a sense of belonging and the loss of a sense of belonging means. Alongside the pride I have in my forebears who worked so hard on the land, it enables me to think about my privilege in a country where land was gained by conquest.

Perhaps what ecopoetry enables me to think most about is the pain and sorrow expressed by Māori during the time when Pākehā voices dominated Aotearoa New Zealand poetry in English. The word 'keening' comes to mind. It is a word I have read twice today in ecopoems by Māori writers. Ecopoetry divulges what was lost and what was not listened to. It has been shocking at times to read ecopoems filled with pain for the loss of land, mana and indigenous plants and species written by marginalised poets at the same time as ecopoems that romanticised colonially constructed landscapes were included in this country's mainstream poetry tradition.

The early years

Ecopoetry by poets born in the nineteenth century or before

> *Poetry is part of the fabric of Māori culture; it is in the oratory; it has been a way to hand down genealogical, historical, and spiritual knowledge, and it is whakapapa ... In Māoridom, we trace our genealogy back all the way to what non-Māori would consider to be inanimate objects – the rivers, the mountains and the land itself. These are my ancestors. Whakapapa is utterly fundamental and in te ao Māori everything begins with, is surrounded by, and originates from whakapapa.*
> —ANAHERA GILDEA (Ngāti Tukorehe)[1]

Intimate knowledge of the attributes of plants and the characteristics and behaviours of birds and other animals is evident in the oral tradition in te reo Māori, where they are often figured as metaphors for human behaviours or as models for people to draw on for knowledge and strength. By tracing whakapapa back to rivers, mountains, the land and the sea, the te reo Māori oral tradition reveals not only a physical and psychic connection with these natural elements but also a seamlessness between the human and what a Western worldview would term the 'nonhuman' worlds. This comprehension of nature and the human relationship with it extends Eurocentric notions of ecology and also, therefore, of ecopoetry.

The traditional pepeha 'Ruia, ruia, tahia, tahia' portrays the long-distance migration of kuaka (godwits) as a metaphor for the benefits of communal support during extended periods of struggle. It was composed by Muriwhenua leader Tūmatahina (c. 1475), who evacuated his people from their besieged pā on Murimotu, the island off North Cape. By travelling hand-over-hand on a rope strung from the island to a cave on the mainland, Tūmatahina saved his people from Ngāpuhi invaders. Merata Kawharu writes that Tūmatahina chanted the pepeha, which ends 'Ko taku ika rā / E whai takoto ana te apunga (There is my band / Formed together into a group)' to embolden his Muriwhenua people against the threat of battle.[2]

> *The words 'ruia' and 'tahia' in this pepeha are used in reference to the preparation of flax for rope; and the 'kuaka' is mentioned because these birds always move in flocks. Thus the saying is now used to show the effectiveness of group action or to stress the importance of bringing people together.*[3]

Merimeri Penfold recalled that the pepeha is also about leadership and the arrival of guests or manuhiri:

> While kuaka fly in flocks, there is one that lands first and then all follow. When visitors are welcomed, they are likened to the godwits or kuaka mārangaranga. Visitors come and go just like the migratory birds.[4]

By linking leadership, group action and the transience of guests, this poem imbues the annual arrival of godwits to Aotearoa New Zealand with cultural expression.

In the chant 'He Waiata Whakaaraara Pā / A Sentinel's Song' composed by Pua-Roro of Ngāti Toa (c. 1670), bird species that inhabit the forest floor warn of an advancing war party:

He kiwi, he weka, he toko kōkako	The kiwi, the weka, the high-stepping kōkako
Kia hara mai hei toko	They come as allies
Mō tō taokete, mō Tara-pu-umeume,	For your brother-in-law, Tara-pu-umeume,
He waewae huruhuru	He of the hairy legs[5]

Jane McRae and Hēni Jacob explain that kiwi and weka:

> are both swift runners as a scout on the lookout needs to be, and the 'high-stepping kōkako' is surely a humorous reflection of the composer treading delicately – like the high-stepping ritual challengers to distinguished visitors to a marae.[6]

In the phrase 'as allies / For your brother-in-law', the kōkako, weka and kiwi are presented as kaitiaki (guardians) whose behaviours are important both as sources of human protection and as models for beneficial human behaviours.[7]

The whakataukī 'Ko te moana, ehara rawa i te wai kau / The sea is not any water' depicts birds and sea creatures as cultural signifiers.[8] Tangaroa, god of the sea and a son of Ranginui, sky father, and Papatūānuku, earth mother, is portrayed in human terms, and the ecological interdependence between sea creatures, birds and people is explicit:

Ko te moana Ehara rawa i te wai kau	The sea is not any water
No Tangaroa ke tena marae	It is the marae of Tangaroa
He maha ona e hua e ora ai	It yields life for many things
nga manu o te rangi	the birds in the sky
te iwi ki te whenua	the people upon the land

Expanding on such ecological connections is the relationship with the sea in which Tangaroa is tūpuna:

> *Māori relationships with taonga in the environment – with landforms, waterways, flora ... and so on – are articulated using kinship concepts. Indeed, the first step in understanding the Māori relationship with the landscape ... is to understand that descent from it is an essential Māori belief.*[9]

In 1889, just nine years before the publication of William Pember Reeves's fatalistic lament 'The Passing of the Forest', Wairarapa leader Pāora Te Pōtangaroa reissued a mōteatea warning of the need for iwi to unite against British acquisition of Māori land.[10] Te Pōtangaroa's waiata is an extension of 'Kaore Hoki Taku Manukanuka', composed c. 1871 by Te Kooti Arikirangi Te Tūruki, which also urged iwi to unite against land confiscation.[11] Te Kooti's and Te Pōtangaroa's versions of the waiata both reference the behaviour of the mātuhi (mātuhituhi) or bush wren, a small insect-eating ground bird.[12] Mātuhi were found in pairs or small family groups and kept in contact with each other by frequent calling:

E mara ma e,	Good people
kia huri mai te taringa	let your ear turn
ki te whakarongo ki te tangi a te Matuhito	listen to the bush wren's cry
A tangi nei, tui, tuia, tuituia	crying here, *'Unite, bond together'*

The mātuhi's appeal to unite invokes whakapapa between Māori and indigenous birds. Hirini Melbourne (Tūhoe, Ngāti Kahungunu) observed that whakapapa is not just about tribal affiliation: 'It's also about relating yourself to other beings sharing the same space. Birds and insects are seen as brothers and sisters through the same progenitor.'[13] Te Kooti's mōteatea begins 'Kaore hoki te manukanuka (How great is my worry)' and relates land loss as 'ki a pū tini tū mana (about great changes harming our mana)'. Te Pōtangaroa calls te tikumu (mountain daisy) and te rau o titapu (kakapo chicks) 'nga taonga whakapaipai o mua (the beautiful treasures of the old days)'. Pounamu is 'the mother-lode' and kotuku, huia, tikumu and toroa are 'nga tohu rangatira o te iwi Maori (prestigious symbols of the Maori)'. He invokes the cries of the mātuhi, huia and wharauroa to call on iwi to unite and beware. This mōteatea is ecopolemic in its protest against the loss of land, plants, animals and mana.

Ecopolemic is also present in the nineteenth-century waiata 'He tangi mo te matenga o ngā kai / Lament for a failed crop' composed by Horomona Hapai (Ngāti Porou). This waiata summons the collection of uncultivated foods – kina, fern-root and pith from the trunks of tree ferns – following the failure of a kūmara crop. The crop failure is explicitly blamed on the arrival of Europeans:[14]

> Kauaka te mara harapaki e rangirangia mai—
> He mate ka rōnaki ki te nui raorao!
> Na te mata parau, nāna i tīwara, tōna hemonga he raorao!
>
> You will not dry by the fire the kneaded, steeped food.
> For this evil has come upon the wide, flat lands!
> The ploughshare cut it up, and the new weeds destroy it!

Margaret Orbell explains that the 'kneaded, steeped food' is kao, 'a preparation made by grating, cooking, and drying kumara ... After the arrival of Europeans, failed crops were sometimes blamed on the use of ploughs, and on new weeds inadvertently introduced by Europeans.'[15] Like British poet John Clare's protest against English farmers taking over communal land in his poem 'The Mores' (1831), 'He tangi mo te matenga o ngā kai / Lament for a failed crop' protests against the adverse effects of private farming on land that was previously communally owned.

Metaphors drawing a close connection between people and birds continue in twentieth-century lyric. In the chant 'Ki Kō, Ki Kō' by Te Māreikura Hori Enoka (c. 1930), the behaviours of indigenous bird species are positive models for human behaviour.[16] The tītī (muttonbird or shearwater), which protects its chicks in secreted underground burrows, is a metaphor for a protective realm that might not be evident at first glance. The constant flitting of the tīwaiwaka (fantail) suggests mouri/mauri (life's force) and warns against danger. Kawekaweā (the long-tailed cuckoo) and pīpī wharauroa (the shining cuckoo), which winter in Pacific islands, are welcomed as a sign that mahuru (spring) and planting time has arrived and that food supplies, often short by the end of winter, will soon increase. The huia's tail feather is evoked as a symbol of cooperation (but the huia's call is not mentioned because by 1930 huia were extinct). 'Tuia, tuia', the constant contact call of the mātuhituhi (bush wren), is a metaphor for togetherness. All these metaphors speak to a worldview of whakapapa and kaitiakitanga, reciprocal duties of guardianship between human and natural worlds.

These few examples of early whakataukī, waiata, whakaaraara pā, pēpeha and mōteatea that have been translated into English gesture to the tradition of ecopoetry in te reo Māori, a tradition that predates the English poetry canon.

Colonial settler poets from Europe, whose poetry begins this country's English poetry canon, conceived the natural world as both a physical resource and a source of solace from the material world. There is an obvious contradiction between these two views of nature. On one hand, nature is valued for what is produced by its destruction; on the other, it is valued for its preservation. This contradiction led to the partitioning of natural areas of land, some of which were confiscated from Māori, into reserves and national parks. Such preservation of natural places was at times assisted by Māori who, under duress, sought to protect taonga, but for Pākehā these areas were generally viewed as scenery. Outside of these reserves, land was quickly, and with great effort, transformed from forest to farmland.

From the late nineteenth century, European settler poets struggled to find an idiom to describe their new terrains. A New Zealand vernacular was developing among those who worked the land and who became known as 'old chums', as opposed to those fresh-off-the-boat 'new chums'. Compared with the florid language of the Victorians, the unpretentious plainness of the old chums' vocabulary was a source of humour to William Pember Reeves and George Phipps Williams. Their 'An Old Chum on New Zealand Scenery' (1889) mocks the 'quaint vocabulary of the hardy pioneer' as unimaginative but also suggests settler pride in those unflattering words derived from the hardships of working, rather than admiring, the land. The generic terms 'gully', 'swamp', 'spur', 'creek' and 'scrub' replaced the Victorian terms 'brook', 'streamlet', 'rill' and 'dell', such that 'you will never, I'm afraid, / Hear a self-respecting bushman call a bush a leafy glade'. This poem portrays the contradictions inherent in early European settler relationships with the land and the difficulties they had in finding a suitable language to describe it. From a Victorian literary perspective, the countryside was a place of beauty, but from the standpoint of the old chum with his boots on the ground, it was a battleground.

Early European settler poets used the language of home, which was generally England, to portray their sense of both the beauty of the local scenery and entrapment in a strange land. In 'To the Makomako, or Bell-bird (Now rapidly dying out of our land)' (1926), Alexander Bathgate uses old-fashioned language to recall the birds of England, 'our Fatherland … whence old mem'ries spring'. His poem mourns yet almost shrugs its shoulders at the expected extinction of a bird species because of colonial deforestation. The makomako or bellbird, now better known as the korimako, did not in fact, 'die out' although its numbers were severely reduced. In 2021, thanks to conservation efforts, its distinctive call was heard in Auckland for the first time in 100 years.

Typically, colonial European poets like Bathgate lamented the wholesale destruction of forests but could envision no other future. Even so, Reeves's 'The Passing of the Forest' (1898) evokes two ideas that were to become ecopoetry tenets. The first is that everything is connected. The poem recognises that the loss of the forest also means the loss of the plant, animal and bird life that the forest sustained: 'Gone is the forest's labyrinth of life, /

Its clambering, thrusting, clasping, throttling race.' The second is the idea that nature benefits human wellbeing: 'Lost is the sense of noiseless sweet escape.' His poem mourns the loss of human comfort that the forest gave.

Like most Pākehā poets of the time, Reeves employed Victorian language and references derived from classical Europe to portray human dominance over a nature perceived as God's beautiful work:

> Scan
> The blackened forest ruined in a night,
>   The sylvan Parthenon that God will plan
> But builds not twice.

Similarly, in 'The Forty-Mile Bush' (1926), Anne Glenny Wilson adopts the style of language that the bushman in 'An Old Chum' scorned: 'the forest's aromatic glade' and 'the murmur of the leafy deep'. When her poem was published, the Forty-Mile Bush in Wairarapa that it so fondly recalls was largely milled and burnt. And by the time Francis Hutchinson's 'Drought' (1950) was published, the prevailing landscape was no longer forest but pasture for grazing livestock. By then, the overwrought Victorian vernacular had been replaced with language the bushman would have been familiar with: 'The grasses wilt and wither, faint and fade ... Parched are the high land water holes, / And far below the creeks shrink fast.'

Blanche Baughan's 'A Bush Section' (1908) employs a language of realism rather than Victorian mawkishness to portray the desolation of land deforested by fire. But the grim landscape of 'prone, grey-black logs ... a silent, skeleton world ... Ruin'd, forlorn, and blank' is nevertheless the site of 'Bright Promise on Poverty's threshold!' Thor, the poem's ten-year-old settler child, offspring of a failed European heritage, looks over a perceived terra nullius, 'this rough and raw prospect, these back-blocks of Being ...' He has the chance to make of it what he will: 'O pioneer Soul! Against Ruin here hardily pitted, / What life wilt thou make of existence?' The train, which rushes through the violated landscape, is 'Gold! gold on the gloom!', a welcome sign of forthcoming industrial progress.

Ursula Bethell shifted away from this Victorian narrative of ruination for improvement to a vision of nature altered by human impact but enduring. When she looks up from her garden in 'Pause' (1929), she recognises that the Canterbury Hills have also been shaped by the weather:

> How grandly the storm-shaped trees are massed in their gorges
> And the rain-worn rocks strewn in magnificent heaps.

Bethell combines realism with Christian terminology to portray a wider vision of colonial impact than the earlier Victorian notions of homogeny and erasure. Her nature is perceived as awe-inspiring compared with temporary human activities, which she sums up succinctly in the final stanza of 'Pause':

> The Mother of all will take charge again,
> And soon wipe away with her elements
> Our small fond human enclosures.

Bethell's rendering of the restorative effects of an enduring nature, as opposed to the Victorian idea of nature as a casualty of human domination, persists in contemporary ecopoetry.

The flip side of the Romantic notion of the beneficial effects of nature on human consciousness is the idea that nature's loss has a detrimental effect on the human psyche. Alan E. Mulgan's 'Dead Timber' (1926) extends the concept of a lost sense of self due to nature's loss, as evoked in British poet John Clare's sonnet 'I am' (1845), by rendering nature's destruction as detrimental to the psyche of the nation as a whole. Mulgan portrays the devastated landscape in Aotearoa New Zealand as the product, or perhaps cause, of a nationhood lacking in creativity: 'There, on the hillside, is our nation's building, / The tall dead trees so bare against the sky.' Rather than developing a cultural heritage, European settlement builds a gallows:

> Yet if some ask: 'Where is your art, your writing
>   By which we know that you have aught to say?'
> We shall reply: 'Yonder, the hill-crest blighting,
>   There is our architecture's blazoned way.
> This monument we fashioned in our winning,
> A gibbet for the beauty we have slain'

This stark assessment evokes the sense of a cultural terra nullius. Pākehā poets writing in the first half of the twentieth century, in what has become known as the nationalist literary period, sought to fill this perceived cultural gap using the landscape as a central trope. During this time, articulating a sense of nationhood and belonging was an important component of much Aotearoa New Zealand ecopoetry in English.

Hutia te rito o te pū harakeke
Kei whea te kōmako e kō?

If you destroy the flax plant
From where will the bellbird sing?

(Traditional)

Titiro kau ana ko ngā pari pōhatu
E whakaatu atu nei i ngā tīpuna.

I look to the rock cliffs
and see the faces
of my ancestors.

(Traditional)

TŪMATAHINA
Ruia, Ruia, Tahia, Tahia

Ruia, ruia, tahia, tahia,
Kia hemo atu te kākoa,
Kia herea mai ki te kawau korokī
E tātaki mai ana
I roto i tana pūkoro whaikaro.
He kuaka mārangaranga,
Kotahi i tau atu, tau atu.
Ko taku ika rā
E whai takoto ana te apunga.

Scatter, scatter,
Sweep on, sweep on,
To pass along on the hard fibres
To be bound to the chatter of the shag
Which is caught inside his protected hollow.
There the restless godwits,
One lands on the sandbank, then another.
There is my band
Formed together into a group.

(c. 1475)

Ko te Moana, Ehara Rawa i te Wai Kau
The Sea is Not Any Water

Ko te moana
Ehara rawa i te wai kau
No Tangaroa ke tena marae
He maha ona e hua e ora ai
nga manu o te rangi
te iwi ki te whenua

The sea is not any water
It is the marae of Tangaroa
It yields life for many things
the birds in the sky
the people upon the land

(Traditional. Shared by Te Ahukaramū Charles Royal, 1989)

PUA-RORO

He Waiata Whakaaraara Pā (A Sentinel's Song)

Te tai rā! te tai whakarongo kī,
Whakarongo kōrero i pū ai te riri,
I mau ia te pakanga.
Nau mai, nau ake!
Kei te tihi, kei te tihi,
Kei te pari, kei te pari,
Kei mata-nuku, kei mata-rangi,
Nohoanga atu o tū ā tāne,
Tēnei nei te para-tāhae,
Whakamataku ana te taringa,
Ko ngā tarutaru e maha,
O te pūkohu o te ngahere,
O Te Wao-nui-a-Tane,
He kiwi, he weka, he toko kōkako,
Kia hara mai hei toko
Mō tō taokete, mō Tara-pu-umeume,
He waewae huruhuru,
Moe ... ! au ... !

The sea yonder, the sea listening for some word,
Listening to the full tale of war,
Which leads on to battle,
Welcome, ascend hither,
At the summit, at the summit
On the cliff-top, on the cliff-top,
On this earth, under the heavens,
Is the eerie post of the young at heart,
Perhaps sneaking hither is another,
His ears a-tingle with fear,
As he treads upon the spreading herbage

Of the mossy woodland floor,
Of the Great-forest-of-Tane.
The kiwi, the weka, the high-stepping kōkako,
They come as allies
For your brother-in-law, Tara-pu-umeume,
He of the hairy legs,
So sleep on! Sleep soundly!

(c. 1670)

HOROMONA HAPAI

He Tangi mo te Matenga o Ngā Kai (Lament for a Failed Crop)

Haere rā, e te kai! E huna i a koe!
Haere rā, e te kai! Te wehi o te tangata
Ki ngā tira haere, i tū ai Ruarua
Ki te poroporoaki:
'Hoatu koe, e te kai, ko au hei muri nei!'

E hine mā e! E ahu ki waho rā
Ki ngā kai tiotio a Hine-ngoingoi, ē,
E kowhete mai nei ki ana tamariki, ē,
Kei tīkina atu hei tami kōmahi,
Pae ana ko te huka i te waha, ē!

E tama mā e! E ahu ki uta rā
Ki ngā kai a Toi, i mahue i muri rā—
Te aruhe, te mamaku, te pou o te tangata ē!

Kauaka te mara harapaki e rangirangia mai—
He mate ka rōnaki ki te nui raorao!
Na te mata parau, nāna i tīwara, tōna hemonga he raorao!

Moumou hanga noa taru nei a te rua!
Tē whakatauria ki ngā kai o Poutū-te-rangi—
Tū tonu i runga, i te whakatatara!

Goodbye, food! Go and hide yourself!
Goodbye, food! From fear of men
In travelling parties, Ruarua stood up
To say farewell:
'Go now, food, and I will follow after!'

O women, make your way to the sea
For the prickly food of the Old Lady.
Who is moaning for her children
Lest they be fetched as a relish for the rotted kumaras
To make our mouths water!

O men, make your way inland
For the foods of Toi, who left behind him
The fern-root and the tree-fern, sustenance for mankind!

You will not dry by the fire the kneaded, steeped food,
For this evil has come upon the wide, flat lands!
The ploughshare cut it up, and the new weeds destroy it!

All these foodpits are quite useless!
Poutu-te-rangi's foods have not come down to them—
He still stands above, far distant!

(c. 1880–1890)

TE KOOTI TE ARIKIRANGI TE TŪRUKI AND PĀORA TE PŌTANGAROA
Kaore Hoki Taku Manukanuka

Kaore hoki e taku manukanuka,
ki aku tini mahara,
e pupuke ake nei i roto i te Hinengaro,
e kore e taea te pehi iho ki roto ra,
me panui atu kia rongo mai
te tini, te mano,
ki te rau e pae nei.
E hoa ma e, katahi ano au,
a Tamairangi,
ka akoako ake ki te tito,
ki te hangahanga rau e.

Ka pere taku pere ki te Tai Rawhiti,
ki a Hinematioro, e kui e,
Tenei tonu au ki te kimikimi ake,
kei te wawae ake i roto
i te ururua
kia ki te au i to hihi o te ra
kia pa mai te mahanatanga ki taku kiri,
kia kohakoha taku noho
kia tae te mahana ki te tau o toku ake,
ka papa piritia, ka haku turitia
i te awhitanga a to mataotao.
Ka pere taku pere
ki te tihi o Hikurangi,
E mara ma e,
kia huri mai te taringa
ki te whakarongo ki te tangi a te Matuhi,
A tangi nei, tui, tuia, tuituia.

E, ka pere taku pere ki te Tai Tokerau,
ki te Tiriti o Waitangi,
E mara ma e, kia huri mai te taringa
ki te whakarongo ki te tangi a te Matuhi,
A tangi nei, 'tui, tuia, tuituia'.

Ehara i te Matuhi na ona tohu
ki te whakamohio,
me aha i te Pukenga,
te pakoko tawhito a Te Riakiaki,
nga punga mahara
o tua iho,
o tua whakarere rawa ei.

E, ka pere taku pere
ki te Tai Tuauru,
kia Mahinarangi, e pa ma e,
ko te purapura tenei a Potatau,
i whiua mai
ki te upoko o te motu nei,
Mahau e ui mai i te aha tonu i,
to matau roa nei,
ka whakahemo nei nga tangata
ki te po,
ka pau nei ki te Reinga.

Maku hoki ia e ki atu, e tai ma e,
he ururua, he titohea ao te oneone,
i te roa hoki ki te whakatiputipu,
ki te rauhi i te taranata i mauria mai nei
e Rakai-hiku-roa
ki te upoko o te motu nei.

E, ka pere taku pere
ki te tihi o Taranaki,
e Whiti ma e,
tenei ano au kei te whakahou
i te ohaki a te tangata
kua mate ki te po.
E hoa ma e, kia huri mai te taringa
ki te whakarongo ki te tangi a te Matuhi,
e tangi nei tui, tuia, tuituia.

E, ka pere taku pere
ki te nuku o te whenua
ki titia taku pere ki te tihi o Tongariro,
kia Hinana ki uta, kia Hinana ki tai.
E hoa ma e,
kati hoki ra te Hinana ki tai ra,
Hinana iho ki uta ra.
Tirohia iho nga manu mohio e toru,
e korero nei i runga i a Aotearoa,
e tangi nei hoki te Matuhi
'Tui, tuia, tuituia'.

E, ka pere taku pere,
ka kau i te tuahiwi o Raukawa,
e ka titia, taku pere
ki te tihi o Tapuaenuku
kia Taiaroa
kia Tuhawaiki
kia Tamaiharoa
e tama ma kia huri mai te taringa
ki te whakarongo ki te tangi a te Matuhi,
e tangi nei, tui, tuia, tuituia.

E rauna noa te Waipounamu,
te wahi i takoto ai
te kuru-auhunga,
te kuru-tongare rewa,
te tiki pounamu,
te taramea
te tikumu,
te rau o titapu,
nga taonga whakapaipai o mua.

Nga tohu rangatira o te iwi Maori,
titia ki runga te upoko
o te piki-kotuku,
te piki-huia;
te raukura,
te tikumu,
whakaheitia ki te taringa,
te pohoi toroa.

Te kuru-ahunga,
te kuru-tongarerewa,
heia ki te kaki
te hei taramea,
te tiki Pounamu, hei aha,
hei whakapaipai ra,
hei whakata kunekune.

Kia pai ai, kia huro ai,
kia mate mai ai
nga tamahine karea
roto o Aotearoa.

E tangi mai nei hoki te Matuhi
'Tui, tuia, tuituia'
e whakakoia nei hoki te Huia
'Koia, koia'
e tangi mai nei hoki te Huia,
'Hui, huihuia'
e tangi mai ana te Pipi-Wharauroa,
whiti, whiti,
whitiwhiti ora,
na mo nga tau ohinawa.

How great is my worry,
adding to my many anxieties
about what is rising up here in my mind
it can't be kept bottled up inside there
it must be communicated
to many, to thousands,
to the bounds of the horizon.
O friends, not long ago
Tamairangi
taught me to spot falsehoods,
by gathering and comparing them.

I will fling my dart towards the East Coast,
towards Hinemaioro, ma'am,
Here I'll keep on checking the falsehoods
So as to reach 'the dead centre
of the undergrowth,'
so that I may behold the sun's rays
and have it bring warmth to my skin,
then, when I vacate my place here
its warmth will reach my friends
sticking flat, making you sweat
and curing your coldness.

Yes, I will fling my dart
at the summit of Hikurangi
Good people
let your ear turn
and listen to the bush wren's cry,
crying here, *'Unite, bond together.'*

Next, I will fling my dart towards the North,
towards the Treaty of Waitangi,
Good people, let your ear turn
to listen to the bush wren's cry,
crying here, *'Unite, bond together.'*

It isn't the bush wren warnings
that should be the notification
for the Bay of Plenty people,
but the ancient carving of Te Riaki
and the long-remembered anchor stones
of yours down below
then eventually abandoned.

I will fling my dart
towards the Kawhia coast
under Mahinaarangi's patronage;
this seed, by Potatau
has already been presented
to the head (government) of this land.
Nevertheless it is for you to untangle things
with your extensive knowledge,
on behalf of the people gone away
to the world of darkness
all fading away at Cape Reinga.

I shall go back even further, oh friends,
scrubby and barren was the soil
it was a long job to return it to fertility
with care and talent, brought about by
you East Coast people
to the government.

Next, I'll fling my dart
towards the summit of Taranaki
to Whiti and his followers
I am still revising
the final speech of that body
which is dying in the night.
My friends, let your ear turn
to listen to the bush wren's cry,
to this cry, '*Unite, bond together.*'

Then I will fling my dart
towards the remote part of the country
to stick my dart onto the summit of Tongariro
in order to 'Fiercely search the land and sea.'
Oh, friends,
blocked off now is our roving on the seas
and soon, our access to the hinterlands.
The three wise birds will soon be observed
speaking here all across New Zealand.
Cry this message again bush wren
'*Unite, bond together.*'

Now, fly my dart,
swim from the shallows of Cook Strait
and stick, oh my dart,
on the top of the Kaikoura ranges,
on the Otago Peninsula,
in Southland,

and Te Maiaharoa's Waitaki basin lands.
O young men there, you turn your ears
to listen to the cry of the bush wren,
to this cry, *'Unite, bond together.'*

Wander freely around the South Island
a bit further to where there is
an abundance of pale green jade
large quantities of semi-transparent jade
the tiki-grade dark green jade
the scent-providing spear grass
the mountain daisy
lots of kakapo chicks
the beautiful treasures of the old days.

The prestigious symbols of the Maori people
stuck on the head of
the devotee of the white heron
and the devotee of the huia;
the plume of feathers
the mountain daisy,
or the albatross-feather ornament
to be attached to the ear.

The far south is the mother-lode of
the most precious things
hung around the neck;
the sachet of scent
the greenstone tiki etc, hanging there
hanging so beautifully
resting plumply.

Such good times, so happy
will soon be coming to an end
for our lovely young women
all around Aotearoa.

And once again the bush wren cries
'Unite, bond together.'
while the huia bird agrees
'Indeed, indeed!'
and the huia also cries out,
'Gather to discuss problems'
while the shining cuckoo cries out
'Beware, beware;
make changes in your lives,
to protect your long-term development.'

(c. 1871–1881)

ALEXANDER BATHGATE

To the Makomako, or Bell-bird
(Now rapidly dying out of our land)

Merry chimer, merry chimer,
  Oh, sing once more,
  Again outpour,
Like some long-applauded mimer,
  All thy vocal store.

Alas! we now but seldom hear
  Thy rich, full note
  Around us float,
For thou seem'st doomed to disappear,
  E'en from woods remote.

Some say the stranger honey-bee,
  By white men brought,
  This ill hath wrought;
It steals the honey from the tree,
  And it leaves thee naught.

The songsters of our Fatherland
  We hither bring,
  And here they sing,
Reminding of that distant strand
  Whence old mem'ries spring.

But as the old we love the new;
  Fain we'd retain
  Thy chiming strain,
Thy purple throat and olive hue:
  Yet we wish in vain.

Thy doom is fixed by Nature's law;
   Why, none can tell.
   Therefore farewell;
We'll miss thy voice from leafy shaw,
   Living silver bell.

Why should we ever know new joys,
   If thus they pass?
   Leaving, alas!
Wistful regret, which much alloys
   All that man now has.

(1926)

ANNE GLENNY WILSON

The Forty-Mile Bush

Far through the forest's aromatic glade
   We rode one afternoon of golden ease.
The long road ran through sunshine and through shade,
   Lulled by the somnolent stories of the trees.

Sometimes a bell-bird fluted far away;
   Sometimes the murmur of the leafy deep,
Rising and falling through the autumnal day,
   Sang louder on the hills, then sank to sleep.

Before us stretched the pine-trees' sombre miles,
   Soft lay the moss, like furs upon the floor;
Behind, the woodland's green monotonous aisles,
   Closed far away by sunset's amber door.

League after league the same. The sky grew red,
   And through the trees appeared a snowy gleam
Of lonely peak and spectral mountain-head,
   And gulfs that nurse the glacier and the stream.

Deep in the glen, the merry waters racing
   Sent forth their turbulent voices to the night;
The stars above began their solemn pacing,
   And home-like shone the distant village light.

Mysterious forest! In this humming city
   I seem to hear thy music-breathing tree,
Thy branches wave and beckon me in pity,
   To seek again thy hospitality!

(1926)

GEORGE PHIPPS WILLIAMS AND WILLIAM PEMBER REEVES

An Old Chum on New Zealand Scenery

I believe it is acknowledged, as a fact, on every hand,
That our own adopted country is a most enchanting land;
And its friends maintain with raptures, which they find it hard to curb,
That its climate's quite unequalled, and its scenery superb.
But the ordinary settler cannot easily express
His ideas on the subject of this wondrous loveliness;
And suppose you ask for details of the beauties that he sees,
I'm afraid that his description may not altogether please—

I have lived among the ranges, on a rough back-country run;
Seen the weather's frequent changes, feed all scorched by summer sun,
Floods in all the snowy rivers, floods in all the rainy creeks,
I have seen the snow in winter wrap the country up for weeks:
I am used to hot Nor'-westers, I am used to wet ones, too,
And the cold Sou'-wester knows me; it has often soaked me through;

I have forded all the rivers, swum them when I knew I must,
After many wet encounters I regard them with distrust;
I have mustered all the country, driven sheep on ev'ry track,
Watched the dogs in ev'ry gully hold them fast or head them back;
I know all the lowest saddles, ev'ry terrace, ev'ry pass,
And the carrying capacity of all the native grass;

I am just a simple squatter, mortgaged to the hilt, I own,
And I boast an honest title to the scenery alone.
I am surfeited with mountains, terrace, gully, spur, and hill,
I have gorged myself with gorges, and of creeks I've had my fill—
So if knowledge of the subject could avail the tale to tell,
I should paint in glowing language all the scenes I know so well.

WILLIAM PEMBER REEVES

The Passing of the Forest
A lament for the children of Tane

All glory cannot vanish from the hills.
   Their strength remains, their stature of command
O'er shadowy valleys that cool twilight fills
   For wanderers weary in a faded land;
Refreshed when rain-clouds swell a thousand rills,
   Ancient of days in green old age they stand,
Though lost the beauty that became Man's prey
When from their flanks he stripped the woods away.

But thin their vesture now—the trembling grass
   Shivering and yielding as the breeze goes by,
Catching quick gleams and scudding shades that pass,
   As running seas reflect a windy sky.
A kinglier garb their forest raiment was
   From crown to feet that clothed them royally,
Shielding the secrets of their streams from day
Ere the deep, sheltering woods were hewn away.

Well may these brooding, mutilated kings,
   Stripped of the robes that ages weaved, discrowned,
Draw down the clouds with soft-enfolding wings
   And white, aerial fleece to wrap them round,
To hide the scars that every season brings,
   The fire's black smirch, the landslip's gaping wound,
Well may they shroud their heads in mantle grey
Since from their brows the leaves were plucked away!

Gone is the forest's labyrinth of life,
   Its clambering, thrusting, clasping, throttling race,
Creeper with creeper, bush with bush at strife,
   Struggling in silence for a breathing space;

Below, a realm with tangled rankness rife,
   Aloft, tree columns in victorious grace.
Gone the dumb hosts in warfare dim; none stay;
Dense brake and stately trunk have passed away.

Gone are those gentle forest-haunting things,
   Eaters of honey, honey-sweet in song,
The tui and the bell-bird—he who rings
   That brief, rich music we would fain prolong,
Gone the woodpigeon's sudden whirr of wings,
   The daring robin all unused to wrong,
Ay, all the friendly friendless creatures. They
Lived with their trees and died and passed away.

Gone are the flowers. The kowhai like ripe corn,
   The frail convolvulus, a day-dream white,
And dim-hued passion-flower for shadows born,
   Wan orchids strange as ghosts of tropic night;
The blood-red rata strangling trees forlorn
   Or with exultant scarlet fiery-bright
Painting the sombre gorges, and that fay
The starry clematis are all away!

Lost is the resinous, sharp scent of pines,
   Of wood fresh cut, clean-smelling for the hearth,
Of smoke from burning logs in wavering lines
   Softening the air with blue, of brown, damp earth
And dead trunks fallen among coiling vines,
   Slow-mouldering, moss-coated. Round the girth
Of the green land the wind brought vale and bay
Fragrance far-borne now faded all away.

Lost is the sense of noiseless sweet escape
   From dust of stony plain, from sun and gale,
When the feet tread where quiet shadows drape

Dark stems with peace beneath a kindly veil.
No more the pleasant rustlings stir each shape,
   Creeping with whisperings that rise and fail
Through glimmering lace-work lit by chequered play
Of light that danced on moss now burned away.

Gone are the forest tracks where oft we rode
   Under the silvery fern fronds, climbing slow
Through long green tunnels, while hot noontide glowed
   And glittered on the tree-tops far below.
There in the stillness of the mountain road
   We just could hear the valley river flow
With dreamy murmur through the slumbering day
Lulling the dark-browed woods now passed away.

Fanned by the dry, faint air that lightly blew
   We watched the shining gulfs in noonday sleep
Quivering between tall cliffs that taller grew
   Above the unseen torrent calling deep,
Till like a sword cleaving the foliage through
   The waterfall flashed foaming down the steep,
White, living water, cooling with its spray
Fresh plumes of curling fern now scorched away.

The axe bites deep. The rushing fire streams bright;
   Swift, beautiful and fierce it speeds for Man,
Nature's rough-handed foeman, keen to smite
   And mar the loveliness of ages. Scan
The blackened forest ruined in a night,
   The sylvan Parthenon that God will plan
But builds not twice. Ah, bitter price to pay
For Man's dominion—beauty swept away!

(1898)

BLANCHE BAUGHAN

A Bush Section

Logs, at the door, by the fence; logs, broadcast over the paddock;
Sprawling in motionless thousands away down the green of the gully,
Logs, grey-black. And the opposite rampart of ridges
Bristles against the sky, all the tawny, tumultuous landscape
Is stuck, and prickled, and spiked with the standing black and grey splinters,
Strewn, all over its hollows and hills, with the long, prone, grey-black logs.

      For along the paddock, and down the gully,
      Over the multitudinous ridges,
      Through valley and spur,
      Fire has been!
  Ay, the Fire went through and the Bush has departed,
  The green Bush departed, green Clearing is not yet come.
      'Tis a silent, skeleton world;
      Dead, and not yet re-born,
      Made, unmade, and scarcely as yet in the making;
      Ruin'd, forlorn, and blank.

At the little raw farm on the edge of the desolate hillside,
Perch'd on the brink, overlooking the desolate valley,
To-night, now the milking is finish'd, and all the calves fed,
The kindling all split, and the dishes all wash'd after supper:
Thorold von Reden, the last of a long line of nobles,
Little 'Thor Rayden,' the twice-orphan'd son of a drunkard,
Dependent on strangers, the taciturn, grave ten-year-old,
Stands and looks from the garden of cabbage and larkspur, looks over
The one little stump-spotted rye-patch, so gratefully green,
Out, on this desert of logs, on this dead disconsolate ocean
Of billows arrested, of currents stay'd, that never awake and flow.
Day after day,
The hills stand out on the sky,
The splinters stand on the hills,

In the paddock the logs lie prone.
The prone logs never arise,
The erect ones never grow green,
Leaves never rustle, the birds went away with the Bush,—
There is no change, nothing stirs!
And to-night there is no change;
All is mute, monotonous, stark;
In the whole wide sweep round the low little hut of the settler
No life to be seen; nothing stirs.

      Yet, see! past the cow-bails,
        Down, deep in the gully,
       What glimmers? What silver
        Streaks the grey dusk?
'Tis the River, the River! Ah, gladly Thor thinks of the River,
His playmate, his comrade,
Down there all day,
All the long day, betwixt lumber and cumber,
Sparkling and singing;
Lively glancing, adventurously speeding,
Busy and bright as a needle in knitting
Running in, running out, running over and under
The logs that bridge it, the logs that block it,
The logs that helplessly trail in its waters,
The jamm'd-up jetsam, the rooted snags.
Twigs of *konini*, bronze leaf-boats of wineberry
Launch'd in the River, they also will run with it,
They cannot stop themselves, twisting and twirling
They too will keep running, away and away.
Yes; for on runs the River, it presses, it passes
On—by the fence, by the bails, by the landslip, away down the gully,
On, ever onward and on!
The hills remain, the logs and the gully remain,
Changeless as ever, and still;
But the River changes, the River passes.

Nothing else stirring about it,
It stirs, it is quick, 'tis alive!
    'What is the River, the running River?
    Where does it come from?
    Where does it go?'

        Listen! Listen ! ...
Far away, down the voiceless valley,
Thro' league-long spaces of empty air,
A sound! as of thunder.
    Look! ah, look!
    Yonder, deep in the clear dark distance,
    At the foot of the shaggy, snow-hooded ranges,—
    Out on the houseless and homeless country
    Suddenly issuing, eddying, volleying—
Smoke, bright smoke! Not the soft blue vapour
By day, in the paddock there, wreathing and wavering,
O'er the red spark well at work in the stumps:
Not the poor little misty pale pillar
Here straggling up, close at hand, from the crazy tin chimney:—
No! but an airy river of riches,
Irrepressibly billowing, volume on volume
Rolling, unrolling, tempestuously tossing,
Ah! like the glorious hair of some else-invisible Angel
Rushing splendidly forth in the darkness—
Gold! gold on the gloom!
... Floating, fleeing, flying ...
Thor catches his breath ... Ah, flown!
Gone! Yes, the torrent of glory,
The Voice and the Vision are gone—
For over the viaduct, out of the valley,
It is gone, the wonderful Train!
Gone, yet still going on: on: on! to the far-away township
(Ten miles off, down the track, and the mud of the metal-less roadway:
Seen, once at Christmas, and once on a fine summer Sunday:

Always a dream, with its dozens of passing people,
Its three beneficent stores) ...
And past the township, and on!
—The hills and the gully remain;
One day is just like another;
In the paddock the logs lie still;
But the Train is not still; every evening it sparkles out, streams by and goes.
>'What is the Train, that it travels?
>Where does it come from?
>Where does it go?'

It is gone. And the evening deepens.
Darker the grey air grows.
From the black of the gully, the gleam of the River is gone.
Scarcely the ridges show to the sky-line,
Now, their disconsolate fringe;
But, bright to the deepening sky,
The Stars creep silently out.
'Oh, where do you hide in the day?'
... It is stiller than ever; the wind has fallen.
The moist air brings,
To mix with the spicy breath of the young break-wind macrocarpa,
Wafts of the acrid, familiar aroma of slowly-smouldering logs.
And, hark, through the empty silence and dimness
Solemnly clear,
Comes the wistful, haunting cry of some lonely, far-away morepork,
'*Kia toa!* Be brave!'
—Night is come.
Now the gully is hidden, the logs and the paddock all hidden.
Brightly the Stars shine out! ...
The sky is a wide black paddock, without any fences,
The Stars are its shining logs;
Here, sparse and single, but yonder, as logg'd-up for burning,
Close in a cluster of light.
And the thin clouds, they are the hills,

They are the spurs of the heavens,
On whose steepnesses scatter'd, the Star-logs silently lie:
Dimm'd as it were by the distance, or maybe in mists of the mountain
Tangled—yet still they brighten, not darken, the thick-strewn slopes!
But see! these hills of the sky
They waver and move! their gullies are drifting, and driving;
Their ridges, uprooted,
Break, wander and flee, they escape! casting careless behind them
Their burdens of brightness, the Stars, that rooted remain.
—No! they do not remain. No! even they cannot be steadfast.
For the curv'd Three (that yonder
So glitter and sparkle
There, over the bails),
This morning, at dawn,
At the start of the milking,
Stood pale on the brink of yon rocky-ledged hill;
And the Cross, o'er the viaduct
Now, then was slanting,
Almost to vanishing, over the snow.
So, the Stars travel, also?
The poor earthly logs, in the wan earthly paddocks,
Never can move, they must stay;
But over the heavenly pastures, the bright, live logs of the heavens
Wander at will, looking down on our paddocks and logs, and pass on.
'O friendly and beautiful Live-Ones!
Coming to us for a little,
Then travelling and passing, while here with our logs we remain,
> What are you? Where do you come from?
> Who are you? Where do you go?'

Ah, little Questioner!
Son of the Burnt Bush;
Straightly pent 'twixt its logs and ridges,
To its narrow round of monotonous labours
Strictly tether'd and tied:

And here to-night, in the holiday twilight,
Conning, counting, and clasping as treasures,
Whatsoever about your unchanging existence
Moves and changes and lives:—
One delight have you miss'd, and that one of more import than any:
More quick than the River, more fraught than the Mail-Train,
More certain to move than the Stars in their courses,
The most radiant wonder, the rarest excitement of all.
  What is it? Oh, what can it be?
  —It is you, little Thor! 'Tis yourself!
  Little, feeble, ignorant, destitute:—
  Wondering, questioning, conscious, alive!
A Mind that moves 'mid the motionless matter:
'Mid the logs, a developing Soul:
From the battle-field bones of a ruin'd epoch,
Life, the Unruin'd, freshly upspringing,
Life, Re-creator of life!

Yea, spark of Life!
Begotten, begetter, of changes:
Yea, morn of Man,
Creature design'd to create:
Offspring of elements all, appointed their captain and ruler:
Here dawning, here sent
To this, thy disconsolate kingdom—
What change, O Changer! wilt thou devise and decree?
Hail to thy god-ship, O Thor! Good luck to the Arm with the Hammer!
Good luck to that little right arm!
Green Bush to the Moa, Burnt Bush to the resolute Settler!
In strenuous years ahead,
Wilt thou wield the axe of the Fire?
Wilt thou harness the horse of the Wind?
Shall not the Sun with his strong hands serve thee, and the tender hands of the Rain?
Daytime and Night spring in turn to thy battle,
Time and Decay run in yoke to thy plough,

And Earth, from the sleep of her sorrow
Waked at thy will, with an eager delight rise, re-quicken'd, and heartily help thee?
—Till the charr'd logs vanish away;
Till the wounds of the land are whole:
Till the skeleton valleys and hills
With greenness and growing, with multiplied being and movement,
Changeful, living, rejoice!

Yea, newly-come Soul!
Here on Earth, from what region unguess'd at?
Here, to this rough and raw prospect, these back-blocks of Being, assign'd—
Lean, cumber'd with ruin, lonely, bristling with hard-ship,
A birthright that fires have been through—
What change, O Changer! creature, Creator, of Spirit!
In this, thy burden'd allotment, wilt thou command and create?
      Finite, yet infinite,
      Tool, yet Employer,
      Of Forces Almighty,
      Beyond thee, within,—
What Fires, of the Spirit, what Storms, wilt thou summon?
What Dews shall avail thee, what Sunbeams? What seed wilt thou sow?
Ease unto weaklings: to thews and to sinews, Achievement!
What pasture, Settler and Sovereign, shall be grazed from the soil-sweetening ashes?
What home be warm in the wild?
Nay, outflowing Heart! thou highway forward and back:
Thought-trains of the Mind! commercing with far-away worlds
What up-country traffic and freight shall travel forth into the world?
What help will ye summon and send?
Spirit, deep in the Dark! with the light of what over-head worlds
Wilt thou in the Dark make friends?
O pioneer Soul! against Ruin here hardily pitted,
What life wilt thou make of existence?
Life! what more Life wilt thou make?

Ah, little Thor!
Here in the night, face to face
With the Burnt Bush within and without thee,
Standing, small and alone
Bright Promise on Poverty's threshold!
        What art thou? Where hast thou come from?
        How far, how far! wilt thou go?

(1908)

DORA WILCOX

In London

When I look out on London's teeming streets,
On grim grey houses, and on leaden skies,
My courage fails me, and my heart grows sick,
And I remember that fair heritage
Barter'd by me for what your London gives.
This is not Nature's city: I am kin
To whatsoever is of free and wild,
And here I pine between these narrow walls,
And London's smoke hides all the stars from me,
Light from mine eyes, and Heaven from my heart.

For in an island of those Southern seas
That lie behind me, guided by the Cross
That looks all night from out our splendid skies,
I know a valley opening to the East.
There, hour by hour, the lazy tide creeps in
Upon the sands I shall not pace again—
Save in a dream,—and, hour by hour, the tide
Creeps lazily out, and I behold it not,
Nor the young moon slow sinking to her rest
Behind the hills; nor yet the dead white trees
Glimmering in the starlight: they are ghosts
Of what has been, and shall be never more.
No, never more!

              Nor shall I hear again
The wind that rises at the dead of night
Suddenly, and sweeps inward from the sea,
Rustling the tussock, nor the wekas' wail
Echoing at evening from the tawny hills.

In that deserted garden that I lov'd,
Day after day, my flowers drop unseen;
And as your Summer slips away in tears,
Spring wakes our lovely Lady of the Bush,
The Kowhai, and she hastes to wrap herself
All in a mantle wrought of living gold;
Then come the birds, who are her worshippers,
To hover round her: tuis swift of wing,
And bell-birds flashing sudden in the sun,
Carolling: ah! what English nightingale,
Heard in the stillness of a summer eve,
From out the shadow of historic elms,
Sings sweeter than our Bell-bird of the Bush?
And Spring is here: now the Veronica,
Our Koromiko, whitens on the cliff,
The honey-sweet Manuka buds, and bursts
In bloom, and the divine Convolvulus,
Most fair and frail of all our forest flowers,
Stars every covert, running riotous.
O quiet valley, opening to the East,
How far from this thy peacefulness am I!
Ah me, how far! and far this stream of Life
From thy clear creek fast falling to the sea!

Yet let me not lament that these things are
In that lov'd country I shall see no more;
All that has been is mine inviolate,
Lock'd in the secret book of memory.
And though I change, my valley knows no change.
And when I look on London's teeming streets,
On grim grey houses, and on leaden skies,
When speech seems but the babble of a crowd,
And music fails me, and my lamp of life
Burns low, and Art, my mistress, turns from me,—
Then do I pass beyond the Gate of Dreams

Into my kingdom, walking unconstrained
By ways familiar under Southern skies;
Nor unaccompanied; the dear dumb things
I lov'd once, have their immortality.
There too is all fulfilment of desire:
In this the valley of my Paradise
I find again lost ideals, dreams too fair
For lasting; there I meet once more mine own
Whom Death has stolen, or Life estranged from me;
And thither, with the coming of the dark,
Thou comest, and the night is full of stars.

(1926)

URSULA BETHELL

Pause

When I am very earnestly digging
I lift my head sometimes, and look at the mountains,
And muse upon them, muscles relaxing.

I think how freely the wild grasses flower there,
How grandly the storm-shaped trees are massed in their gorges
And the rain-worn rocks strewn in magnificent heaps.

Pioneer plants on those uplands find their own footing;
No vigorous growth, there, is an evil weed:
All weathers are salutary.

It is only a little while since this hillside
Lay untrammelled likewise,
Unceasingly swept by transmarine winds.

In a very little while, it may be,
When our impulsive limbs and our superior skulls
Have to the soil restored several ounces of fertiliser,

The Mother of all will take charge again,
And soon wipe away with her elements
Our small fond human enclosures.

(1929)

FRANCIS HUTCHINSON

Drought

The hand of the Sun
Lies heavy on this land.
The solemn drought steals on
The grasses wilt and wither, faint and fade.

First on the high dry terrace-lands,
On grey cliff edges, naked spurs,
The green grows brown and fades to grey.
Parched are the high land water holes,
And far below the creeks shrink fast.
We look to westward, longingly,
But rain so wished-for does not come.

Only the daily portent—
Clouds that, hurrying up, seem full of promise,
Thinning too soon to harsh grey blue
   And boisterous gales.

Is it prophetic impulse that the plants
   Are pushing onward suddenly—
       The great and small alike—
To quick fruition?
The wind bows a myriad bents,
The sward's ablaze with flowers.

(Date unknown)

TE MĀREIKURA HORI ENOKA
Ki Kō, Ki Kō

Ki kō, ki kō—tirohia!
Kei whea te taunga
o te Tītīwaitori?

Parepare mai ra koe
e te Tīwaiwaka
i te paepae
o te tautara
E nono tītaka
te tau
i te mouri,
Kataina mai rā
e te Kōkako,
'*Kōaka, kōaka!*'

Kia whakataukī te manu Tūī,
Tuia, tuia, i te pūaotanga
Kia whakapurua
ki te remu o te Huia.
Ka whakarongo ki te tangi
a te Kawekaweā,
kawea mai ra
i te tō-matomato-tanga
o te tōmairangi.
E rongo koe i te Pīpīwharauroa.
'*Kūī, kūī,*
whitiwhiti ora!'

Kia whakapainga
ki te Manu Tawhiorangi
ka puta,
ka ora nā i!

Over there, over there—look!
Where's the resting place
of the Muttonbird? It's not obvious, eh?

You may be diverted
by the Fantail
from the crossbeam
of the desecrated latrine,
by its flitting about,
constantly alighting
filled with life's energy,
and then be teased
by the Kokako,
'Calabash, empty head!'

And when the Tui bird proclaims
'*Be bound together, together,*' at dawn,
with its throat padded up
like the tail feather of the Huia,
or you listen to the cry
of the Long-tailed Cuckoo
bringing with it
the greening
of the gentle spring dew
or hear the Shining Cuckoo's song
'Short of food, spiritual food,
May your life change for the better!'

In these ways you are being blessed
by the Holy Spirit:
appearing,
and enriching your life in this way!

(c. 1930s)

ALAN E. MULGAN

Dead Timber

These are not ours—the isles of columned whiteness,
   Set in an old and legend-whispering sea;
Nor crowning domes that take the morning's brightness,
   Praising the Lord in open majesty;
Nor arches' hushed eternal invocation;
   Nor windows glowing with the love of God;
Nor slender minarets that take their station,
   Like spears ascending where the faithful trod.

There, on the hillside, is our nation's building,
   The tall dead trees so bare against the sky.
They neither kiss the morn nor take the sunset's gilding;
   They hear no brimming prayer, no sinner's cry.
But in the desolation of our making,
   Where prey at will the sun and wind and rain,
They call the sky to witness of our breaking,
   They tell the stars the story of our gain.

Unranked and formless, stark they stand, unheeding
   The whisper of their brothers, soon to die.
Their hearts are dry from the bright axe's bleeding,
   And dead the music of their leaves' long sigh.
Mute in their misery of devastation,
   They hold between us and the living light,
In twisted agony of revelation,
   The lifeless litter of the field of fight.

Yet if some ask: 'Where is your art, your writing
   By which we know that you have aught to say?'
We shall reply: 'Yonder, the hill-crest blighting,
   There is our architecture's blazoned way.
This monument we fashioned in our winning,

A gibbet for the beauty we have slain;
Behold the flower of our art's beginning,
　　The jewel in the circlet of her reign!'

Yet, so doth patient beauty work, subduing
　　The very husks of death to gracious ends;
The heavy plodding days, their task pursuing,
　　Slowly transmute these victims into friends.
Dwelling with them, we take them to our living;
　　Looking on them, we wed them to our sight.
Resting with us, they grant us their forgiving,
　　And creep into the round of our delight.

Less were the dawn in miracle unfolding,
　　Did these return not to the breathless hill.
Disturbed the heart, known loveliness beholding,
　　Did these not watch us as the hours fill.
Strange were the hush of eve by mists enchanted,
　　Did these not stand to catch the floating flowers,
Common the moonlight by the shadows haunted
　　But for the mystery of these lightless towers.

Some day our feet may walk where art is golden;
　　Then round our hearts will lap the tides of time.
We shall be one with dwellings rich and olden,
　　And fragrant prospects sweet with ancient rhyme.
Yet, though we go where memories come thronging,
　　And wonder leads us wheresoe'er we roam,
Through our delight will creep the voice of longing—
　　O dear dead timber on the hills of home!

(1926)

The middle years

Twentieth-century ecopoetry

During the first half of the twentieth century, the Aotearoa New Zealand poetry canon in English was exclusively comprised of the work of Pākehā poets. A burgeoning sense of belonging to this country and its landscapes was a central theme during what has become known as the nationalist literary period. Ecopoems written during this time brought into question the Victorian headlong rush towards agricultural and industrial development as the negative effects of such rapid so-called improvements became apparent.

By the middle of the century, Pākehā connection with the natural world was evoked through a sense of belonging to the now familiar landscapes constructed by colonialism. Changes in attitudes towards the environment were sparked here as elsewhere during the Cold War period by the global threat of nuclear war and awareness of the downstream effects of chemicals on nature following the publication of American author Rachel Carson's *Silent Spring* (1962). The idea of nature as a limitless resource for human consumption began to shift; those outside the mainstream, at least, started to value nature. This led to the rise of ecological activism that sought political and social change, and in poetry, ecopolemic, calling for ecological action. The Save Manapouri Campaign, which was waged from 1969–72, successfully protested against raising the levels of lakes Manapouri and Te Anau in the South Island as part of the Manapouri hydropower project. The campaign garnered widespread support and encouraged the idea of ecological limits to industrial progress. Along with the 1972 formation of the New Zealand Values Party (which later became the Aotearoa New Zealand Green Party), it signalled a shift in the impetus for environmental protection from activism to mainstream social and political consciousness.

Māori and Pacific poets writing in English and published from the 1950s onwards continued to express embodiment between culture and nature, along with deep sorrow for the loss of land and indigenous species and the loss of mana associated with these absences. With the addition of their voices to the country's English poetry canon, the differences between Indigenous and Pākehā notions of nature and the human relationship with it were revealed more widely. Hone Tuwhare's 'Papa-tu-a-nuku (Earth Mother)', written in support of the 1975 Māori Hīkoi, was a pivotal point in the Māori renaissance. It brought a portrayal of the natural world in human terms to the English poetry canon. Such evocations of cultural embodiment in nature became, as the century unfolded, less and less marginalised, expanding Western notions of ecological encroachment and, therefore, of ecopoetry.

English ecopoetry of the twentieth century begins with a brighter feeling of connection to the landscape shaped by colonialism, contrasting with the grim view of a cultureless nation represented by Alan Mulgan in his 'Dead Timber' (1926). Nevertheless, Allen Curnow's 'House and Land' (1941) reaches back to the notion of isolation from Europe through its portrayal of settler life lived in a 'spirit of exile' from the cultural refinement of Britain. The 'stagnant' weather and a dog 'moping under the bluegums' are a gauge of the 'great gloom' in 'a land of settlers / with never a soul at home.' The poem portrays a deep sense of cultural disconnection between Pākehā settlers, even those born here, and the colonially constructed ecologies they found themselves in. With its introduced bluegums, fowls and rabbits, the poem suggests that the nonhuman world is also lacking in rootedness.

Charles Brasch evoked a sense of coming to terms with those damaged ecologies in 'The Land the People III' (1939). Like Blanche Baughan, he perceived the land as terra nullius, a nobody's land, where 'the newcomer heart' needs 'slow-paced generations ... To make of new air, new earth, part / Of its own rhythm and impetus ...' His poem explains that European settlers will not be at home with this country's landscapes simply by dominating them. 'Only in the wash of time /... can earth and man / Into understanding grow ...' he writes, and:

> Not the conquest and the taming
> Can make this earth ours, and compel
> Here our acceptance.

By moving from a sense of erasure to a notion of adaptation within a constantly transforming environment, Brasch, rather like Ursula Bethell, promotes a sense of European settler belonging not through domination of the nature that remains but rather through connection with it.

The earlier Victorian visions of a landscape dominated by mountains and trees, as depicted by William Pember Reeves, Alexander Bathgate and Anne Glenny Wilson, are now replaced by recognition of the wider transformative effects of agriculture. Ruth Dallas's 'Pioneer Woman with Ferrets' (1976) is a pitiless evocation of life in a land transformed by deforestation and overrun with rabbits.

Farmer and prose writer Herbert Guthrie-Smith's 1921 memoir *Tutira: The Story of a New Zealand Sheep Station* was pivotal in denoting a shift away from lamenting the pre-European nature that was lost through colonisation to a sense of connection with the nature that remained. It recognised that local and recent land use continued to alter ecologies. Peter

Bland's 'Guthrie-Smith at Tutira / *Hawke's Bay–1885*' (1979) echoes Guthrie-Smith's sense of doubt as to whether his life's work of improving land for sheep farming has, in fact, caused its devastation: '[H]ave I then for sixty years desecrated God's earth and dubbed it improvement?'[1] Bland's poem begins, 'Who am I? What am I doing here / alone with 3000 sheep?' It continues, 'I am the one sheep / who must not get lost' and 'knowing this place I learn to know myself', implying the sense of responsibility that Guthrie-Smith came to accept for the ecological harm caused by hill country sheep farming.

Alongside sheep, another introduced species, trout, and the recreational sport of fishing are enduring symbols of Pākehā connection with, and loss of connection from, nature. Brian Turner, one of this country's most prolific conservationist poets, chronicles his passion for both angling and the river environs through a poetic speaker whose connections with Central Otago's trout habitats are portrayed as communions. American author Zane Grey's game fishing guide, *Tales of the Angler's Eldorado New Zealand* (1926), helped establish the Bay of Islands as a premier world game-fishing destination. Grey, a dentist, writer and adventurer whose stories and novels idealised the American frontier, fished in the Bay of Islands in the 1920s and built a lodge on Urupukapuka Island called the Zane Grey Sporting Club. However, another poet and angler, John Newton, found a more equivocal and violent landscape than Grey's fishing paradise. In his 'Eldorado Poem' a trout shares the river with 'the fleece / of a drowned sheep …'

> Duckweed, grey and crimson,
> mirrors the red of the woolshed, the slow
> rot in the yards.

Trout, then, offer both solace and the spectre of ecological damage, a vision expanded on in the next section of this anthology in ecopoems that recall the indigenous fish species that trout supplanted, revealing the complexities of nature in a settler colonial country.

Twentieth-century ecopoetry sees the introduction and influence of poems by Māori and Pacific poets into the country's English poetry tradition. Alistair Te Ariki Campbell was the first Aotearoa New Zealand poet with Polynesian heritage to be published in English with his collection *Mine Eyes Dazzle* (1950). Poems by Te Ariki Campbell, Hone Tuwhare, Rangi T. Harrison, Evelyn Patuawa-Nathan, Vernice Wineera Pere and later in the century, Jacq Carter, Keri Hulme, Trixie Te Arama Menzies, Roma Potiki, Apirana Taylor, Ngahuia Te Awekotuku, Robert Sullivan and Albert Wendt expand on and at times subvert earlier Pākehā comprehensions of ecology and, therefore, of ecological encroachment.

Te Rangi Tanira Harrison's 'Waikato te Awa (Waikato is the River)', translated by Herewini Easton, renders an enduring connection with the river through physical attributes and cultural history. It demonstrates the concept that whakapapa includes landscape. Easton writes:

> Waikato awa begins in the tribal area of Tūwharetoa as a streamlet south of Tūrangi, then flows into Taupōnui a Tia, Lake Taupō. It continues through the tribal boundaries of Ngāti Tahu, Ngāti Raukawa and Tainui Waikato, eventually to Port Waikato, Te Pūaha o Waikato to the Tasman sea, Te Moana Tāpokopoko a Tawhaki.[2]

The poem is written 'in the tempo of pātere, a chant that flows' and 'there are many contexts flowing in from the tupuna awa-river ancestor of Waikato':

> There are whakapapa-genealogies, whanaungatanga-relationships, kai-food resources, wairua-spirituality, whanau-social, hinengaro-emotional and communication, tinana-physicality, mana motuhake-unique characteristics, rangatiratanga-autonomy, kiingitanga-regality and kingship and whai rawa-economical concepts.[3]

Jacq Carter's 'Our Tūpuna Remain' (1998) renders natural elements inhabited by ancestral spirits. Observing farmland, the poem recalls not only lost ecologies but also lost loved ones:

> Nothing like a lone-standing nīkau
> in the middle of some paddock
> owned by some Pākehā
> to make you feel mamae
>
> Surrounded by maunga
> who serve to remind you
> that once the whole paddock
> had that same sense of tapu

Despite 'a change of name / a chopping down of trees / a burning down of whare', Carter asserts 'our tūpuna remain'. Similarly, Roma Potiki's 'A Cloak and Taiaha Journey' (1998) portrays the presence of tūpuna in landforms: 'their unearthly features holding the tatau of land, / sky and water'. She writes: 'They do not speak in the language of sound / but directly to the ones seeing them.' In these poems, natural elements are not only physical features but also sources of cultural expression.

Evelyn Patuawa-Nathan's 'Waikato Lament' (1979) portrays the loss of mana associated with the loss of land through colonisation. 'Green wandering fingers / of kikuyu' (an exotic grass and perhaps a metaphor for European people), are interfering with indigenous ecologies, 'prying into an old kumara pit' yet they 'cannot hide, / the faded emblems / of a land lost people.'

> Blood soaked, in time's
> memory,
> spirits of Taupiri
> raise keening voices
> anthem of injustice
> echoing down
> through the night.

Land was taken by force, and the spirits of ancestors remain in Taupiri mountain where the mana of Waikato iwi is symbolised and centred.

In contrast with the cultural and species losses evoked in Patuawa-Nathan's poem, Tricia Glensor's 'Otago Landscape' (1977) laments the degrading effects of farming. It begins:

> Sheep crawl on the hill's broad back
> like swollen slaters
> under the stone of the sky.
>
> Crushed by the weight of wheels
> the land lies
> stripped and servile.

The barrenness of Glensor's image contrasts with Patuawa-Nathan's poem, which recalls ecological abundance and joy, memories that are absent from Glensor's blunt portrayal of devastation.

The loss of land and consequent loss of traditional food sources through colonisation is extended by Apirana Taylor in 'Feelings and Memories of a Kuia' (1981). The poem is written in the voice of the kuia, who recalls customary food-gathering in precolonial times when bundles of fern drew fish from a lake, and kererū and eels were plentiful in the bush and river. But, she says,

> The bush
> got chopped down
> the river dammed
> and the lake polluted

The consequences of these ecological losses are culturally devastating, leading the kuia to question what Māori sacrificed themselves for in World War II:

> So we could live in quarter-acre sections
> where once we had
> more land
> than my eye could see
>
> So we could eat
> hamburgers
> when once we got
> fresh kai
>
> So in the end
> we could be lost
> and unhappy
> and not know why

Environmental activism was another important influence on the country's ecopoetry during the twentieth century. The 1969–72 Save Manapouri campaign was the catalyst for widespread environmental activism and locally based ecopolemic. In the late 1970s, another successful environmental campaign was launched against a proposal to build an aluminium smelter at Aramoana near Dunedin. Ecopoems by Pākehā and Māori writers reveal different ways of comprehending the smelter's potential for ecological damage.

Ian Wedde's 'Pathway to the Sea' (1975) casts the smelter as an example of global industrial pollution and urges civil responsibility. It maligns corporate power against a generalised nature, 'there's / birds out there', and promotes civic responsibility by addressing the reader as 'citizen'. The final stanza moves beyond local subject matter to a case for global responsibility:

> living in the
>   universe doesn't
>   leave you
> any place to chuck
>   stuff off
>   of.

This poem follows the example of British and American mid-twentieth-century ecopolemic urging individual accountability towards the future protection of the planet in the face of destructive political and industrial power.

In contrast, Hirini Melbourne's 'Aramoana' (date unknown) focuses on the local by naming natural features and resident bird species. Huikaau is the name of the tide that comes through the main channel and also means a gathering of shags. Hautai references the peak on top of the peninsula. Muaūpoko is the Māori name for the Otago peninsula and the albatross colony at the end of the peninsula.[4] The poem's final stanza addresses the albatross and, through the use of the pronoun 'us', connects the threat to its habitat with the colonisation of Māori:

> Albatross of the sky
> Do not land
> There is no place for us
> in this world

Muru Walters's 'Haka: The feathered albatross / He huruhuru toroa' (1985), which also protests against the proposed smelter at Aramoana, further extends the connection between people and nature by looking back to the arrival of the albatross and of Māori from Hawaiki. It asserts Indigenous belonging to 'the ancestral home', which is rendered as a 'confiscated homeland'. The poem thereby connects colonialism with the invasion of both ecology and culture:

> Lofty foreigners! Distant foreigners!
> Aue! Polluting
>   our land
> Look! Look! Look!

By invoking ecological encroachment as the loss of place for indigenous species and Indigenous people alike, 'Haka: The feathered albatross' and 'Aramoana' expand on current definitions of ecopoetry. They portray colonial ecological encroachment, an ecopolemic outside of the Western worldview described by foundational critics of the field.

Campaign for Nuclear Disarmament (NZ) was formed in 1959, and the title poem of Hone Tuwhare's *No Ordinary Sun*, first published in *Te Ao Hou* the same year, evokes the horror and devastation of an atomic bomb. It focuses on the demise of a single tree, which is also a metaphor for humanity, and on the loss of other natural elements, the sea, sun, moon, wind and rain. Tuwhare names the bomb only as 'the monstrous sun', thereby projecting a sense of it as unnatural in a world where nature and people belong. His metaphor of the tree as human, 'Tree let your arms fall', suggests the local rather than the global impact of nuclear apocalypse.

Anti-nuclear ecopolemic continued to be written in the 1990s in response to French nuclear testing in the Pacific. 'Mururoa/Moruroa' (1995) by Ngahuia Te Awekotuku evokes historical cultural connection with the ocean. Beginning with words from The Lord's Prayer, 'Murua o matou hara / Forgive us our trespasses', the poem portrays the French as trespassers in a region inhabited for centuries by Pacific people:

> So the saltwarm of your
> waters move
> ancient in our blood
> remembering
> canoes seeded cast
> a thousand years ago
> from homeland islands

By comparing French nuclear testing with 'the french clap', which was brought to the Pacific Islands by European sailors, the poem frames nuclear testing alongside the historical invasions of colonialism. Keri Hulme also recalls the historical connection of Pacific people with the sea in 'Te Rapa, Te Tuhi, Me Te Uira (or Playing with Fire)' (1995):

> Yes, we are small people from little islands
> but we have sailed our hopes on the age-old path, that
>   wideness of life,
> our birth and making zone, Te Moananui a Kiwa—

She imagines local people and ecologies raising a curse against those 'hoons' who carry out nuclear testing in their homeland. 'We are sea-people', she writes, 'you foul our home.' Hulme portrays the attitude of the earth to nuclear tests: 'Papatūānuku shudders under her mantle of ocean', thereby framing the earth as a part of the human realm. By drawing on cultural responses to nuclear threats, these ecopoems portray ecological encroachments that are specific to Aotearoa New Zealand.

As Pacific and other settler voices in English increase in prevalence in ecopoetry in the twentieth century, so does the inclusion of Māori, Pacific and Asian words, phrases and place names into English poetry. A bilingual, and at times multilingual, approach to ecopoetry continues and grows, particularly as non-Māori New Zealanders become increasingly familiar with some Māori words, and te reo Māori becomes a first language for many tangata whenua. Representation of the connections between Māori, Pacific people and the wider Asia Pacific region is another touchstone of Aotearoa New Zealand ecopoetry in the twenty-first century.

FLEUR ADCOCK

Last Song

Goodbye, sweet symmetry. Goodbye, sweet world
of mirror-images and matching halves,
where animals have usually four legs
and people nearly always two;
where birds and bats and butterflies and bees
have balanced wings, and even flies
can fly straight if they try. Goodbye
to one-a-side for eyes and ears and arms
and breasts and balls and shoulder-blades
and hands; goodbye to the straight line
drawn down the central spine,
making us double in a world
where oddness is acceptable only
under the sea, for the lop-sided lobster,
the wonky oyster, the creepily rotated
flatfish with both eyes over one gill;
goodbye to the sweet certitudes of our
mammalian order, where to be
born with one eye or three thumbs
points to not being human. It will come.

In the next world, when this one's gone skew-whiff,
we shall be algae or lichen, things
we've hardly even needed to pronounce.
If the flounder still exists it will be king.

JAMES K. BAXTER

Waipatiki Beach

1
Under rough kingly walls the black-and-white
Sandpiper treads on stilts the edges

Of the lagoon, whose cry is like
A creaking door. We came across the ridges

By a bad road, banging in second gear,
Into the only world I love:

This wilderness. Through the noon light rambling clear
Foals and heifers in the green paddocks move.

The sun is a shepherd. Once I would have wanted
The touch of flesh to cap and seal my joy,

Not yet having sorted it out. Bare earth, bare sea,
Without fingers crack open the hard ribs of the dead.

2
If anyone, I'd say the oldest Venus
Too early for the books, ubiquitous,

The manifold mother to whom my poems go
Like ladders down—at the mouth of the gully

She had left a lip of sand for the coarse grass to grow,
Also the very quiet native bee

Loading his pollen bags. We parked the car
There, and walked on

Down to the bank of the creek, where the water ran under
A froth of floating sticks and pumice stone,

And saw in the dune's clasp the burnt black
Trunk of a totara the sea had rolled back.

3
Her lion face, the skull-brown Hekate
Ruling my blood since I was born,

I had not found it yet. I and my son
Went past the hundred-headed cabbage tree

At the end of the beach, barefooted, in danger of
Stones falling from the overhang, and came

On a bay too small to have a name
Where flax grew wild on the shoulder of the bluff

And a waterfall was weeping. A sheep leapt and stood
Bleating at us beyond a tangle of driftwood

And broken planks. Behind us floated in the broad noon
Sky that female ghost, the daylight moon.

4
A leper's anger in the moon's disk, or
The long-tongued breaker choked by sand,

Spell out my years like Pharaoh's wheat and husk—
I walk and look for shelter from the wind

Where many feet have trodden
Till silence rises and the beach is hidden.

PETER BLAND

Beginnings

Guthrie-Smith at Tutira
Hawke's Bay–1885

Who am I? What am I doing here
alone with 3000 sheep? I'm
turning their bones into grass. Later
I'll turn grass back into sheep.
I buy only the old and the lame.
They eat anything—bush, bracken, gorse.
Dead, they melt into one green fleece.

Who am I? I know the Lord's my shepherd
as I am theirs—but this
is the 19th century; Darwin
is God's first mate. I must keep
my own log, full of facts if not love.
I own 10,000 acres and one dark lake.
On the seventh day those jaws don't stop.

Who am I? I am the one sheep
who must not get lost. So
I name names—rocks, flowers, fish:
knowing this place I learn to know myself.
I survive. The land becomes
my meat and tallow, I light my own lamps.
I hold back the dark with the blood of my lambs.

ARAPERA HINEIRA KAA BLANK

Conversation with a Ghost 1974–1985

(On looking at other people's houses)

E ki, e ki, waiho o maunga,
Hei paepae koiwi, whare tipuna!
Kei hea a Tamaki-makaurau
Kua riro nei a Maungawhau, Maungarei, Orakei!

I built my home in shadows deep
beneath the mountains
close to where the river rippled
glimmered gave life to earth and me,
I did as I was told!
My bones now ache with
fish-hook gripping pain
from endless damp
Mother earth no longer gleams
with kumara vine to the water's edge.
Sewage reeks where manuka grew,
Tutae laps on every shore
And kutai beds grow fat, alone,
Like worms in burial places.
Where is my mana now
When strangers strangle living bones?

On Maungawhau, Maungarei, Orakei
padlocked in by high-rises boxes
soft-silk, teak-lined, richly furnished
concrete structures, living tombstones?

Dear departed ghost, I ask you now,
What have you done to me
who followed the KAWA,

shared my mana with other people
who wonder why I,
the tangata-whenua,
chooses to live in shadows deep?
For my bones no longer sing!

CHARLES BRASCH

The Land and the People (III)

There are no dead in this land,
No personal sweetness in its earth;
Mountain and forest stand
Solemn and dumb as the forever
Stars, untouched by the sheep's path,
The climbing hand upon the rock
Loverlike, or the watching lover
Humble from far off. And the newcomer heart,
Needing slow-paced generations, the shock
Of recognition after long heedlessness,
Routine and ripening memory,
To make of new air, new earth, part
Of its own rhythm and impetus,
Moves gauchely still, half alien.
Only in the wash of time
Identifying, as the sea
Isolates, can earth and man
Into understanding grow
And to a common instinct come.
Not the conquest and the taming
Can make this earth ours, and compel
Here our acceptance. Dearest dust and shadow
Must we offer still, becoming
Richer as our loss falls home
Into her safer present keeping, who
Compounds our ash with the trees' blood,
The living and the dead inseparable.

ALISTAIR TE ARIKI CAMPBELL

The Return

And again I see the long pouring headland,
And smoking coast with the sea high on the rocks,
The gulls flung from the sea, and dark wooded hills
Swarming with mist, and mist low on the sea.

And on the surf-loud beach the long spent hulks,
The mats and splintered masts, and fires kindled
On the wet sand, and men moving between the fires,
Standing or crouching with backs to the sea;

Their heads finely shrunken to a skull, small
And delicate, with small black rounded beaks;
Their antique bird-like chatter bringing to mind
Wild locusts, bees, and trees filled with wild honey;

And sweet as incense-clouds, the smoke rising, the fire
Spitting with spots of rain, and mist low with rain;
Their great eyes glowing, their rain-jewelled, leaf-green
Bodies leaning and talking with the sea behind them,

Plant gods, tree gods, gods of the middle world. Face downward
And in a small creek-mouth all unperceived,
The drowned Dionysus, sand in his eyes and mouth,
In the dim tide lolling: beautiful, and with the last harsh

Glare of divinity from lip and broad brow ebbing ...
The long-awaited. And the gulls passing over with shrill cries;
And the fires going out on the thundering sand;
And the mist, and the mist moving over the land.

JACQ CARTER

Our Tūpuna Remain

Nothing like a lone-standing nīkau
in the middle of some paddock
owned by some Pākehā
to make you feel mamae

Surrounded by maunga
who serve to remind you
that once that whole paddock
had that same sense of tapu

It's a bit like that urupā
in the middle of that reserve
which used to be a papakāinga
till some Pākehā had it burned

So consider yourselves warned:

It'll take more

than
a change of name
a chopping down of trees
a burning down of whare

to make us forget

our tūpuna remain

ALLEN CURNOW

House and Land

Wasn't this the site, asked the historian,
Of the original homestead?
Couldn't tell you, said the cowman;
I just live here, he said,
Working for old Miss Wilson
Since the old man's been dead.

Moping under the bluegums
The dog trailed his chain
From the privy as far as the fowlhouse
And back to the privy again,
Feeling the stagnant afternoon
Quicken with the smell of rain.

There sat old Miss Wilson,
With her pictures on the wall,
The baronet uncle, mother's side,
And one she called The Hall;
Taking tea from a silver pot
For fear the house might fall.

She's all of eighty said the cowman
Down at the milking-shed.
I'm leaving here next winter,
Too bloody quiet he said.

The spirit of exile, wrote the historian,
Is strong in the people still.
He reminds me rather, said Miss Wilson,
Of Harriet's youngest, Will.

The cowman, home from the shed, went drinking
With the rabbiter home from the hill.

The sensitive nor'-west afternoon
Collapsed, and the rain came;
The dog crept into his barrel,
Looking lost and lame.
But you can't attribute to either
Awareness of what great gloom
Stands in a land of settlers
With never a soul at home.

RUTH DALLAS

Pioneer Woman with Ferrets

Preserved in film,
As under glass,
Her waist nipped in,
Skirt and sleeves
To ankle, wrist,
Voluminous
In the wind,
Hat to protect
Her Victorian complexion,
Large in the tussock
She looms,
Startling as a moa.
Unfocussed,
Her children
Fasten wire-netting
Round close-set warrens,
And savage grasses
That bristle in a beard
From the rabbit-bitten hills.
She is monumental
In the treeless landscape,
Nonchalantly she swings
In her left hand
A rabbit,
Bloodynose down,
In her right hand a club.

HARRY DANSEY

The Old Place

This is the place where the old people
lived. They caught the birds, stored
them in their own rich fat, grubbed
fern root, loved, mated, buried
their dead in the rocks and crannies
and on the high cold hill.

Here came Uenuku, broke the tapu
of the chief's spring, left his
deed in a proverb. Here the
old man hauled a tōtara, with
his own hands hewed a ridge-pole
fifty feet from the sound red heart.

There by the alien pines his house
stood, silver-grey in its dotage,
and his church there where the
six-foot fern sways brown and dusty;
all vanished in the scrub fires
in the years when no one cared.

So I park the landrover, climb
the slope, push aside the broom,
hope for a sign from the past
from the old dead people,
but there is no more comfort here
in the fierce bright silence
than the rasping tūī finds in
black bark and the hard pine needles.

LAURIS EDMOND

Atom Bomb Test, Moruroa Atoll, 6 September 1995

I am water, I am sand
I am a cell in the trembling earth
I am a shaken pebble on the hurt sea floor

a young fish made ill by the predator poison
coursing towards me across the ocean
that was my friend;

I am the child's brown toes
curling back from the tumult of gravel
at tide's edge grown suddenly dangerous and sharp

I am the hand, the foot, the easily bleeding veins
the skin that a monstrous tearing apart
of the air may lift from the flesh;

I am the child beaten to blindness by a flash in the sky
the orphan staggering about my dark; I am
the fear of every creature for grotesque adventures

of brilliant men. Call to me in my fear,
cry for me, take me from the tainted earth
that was my home, the ground where in an instant

the innermost cell of life may have no place,
tell me what I must do when the simplest act of living
is undone and turned to chaos.

TRICIA GLENSOR

Otago Landscape

Sheep crawl on the hill's broad back
like swollen slaters
under the stone of the sky.

Crushed by the weight of wheels
the land lies
stripped and servile.
Fetters of thorn and barbed wire
force her to kneel
and feed us.

Saddled by laws,
we walk within our hedges;
feel the eggs,
which the years laid under our skin,
hatch into leeches.

DENIS GLOVER

Lake Manapouri

For a million years
Or some such improbable time
The lake kept itself to itself,
A policy of isolation.
Bush and mountains generated no myths.

No Lady of the Lake was there to dust
Those myriad isles (twenty-nine anyway),
Besom that bosomed floor.

Incurious wildlife saw
In all that water one big birdbath;
Narcissus mountains powdered ballet white
Adored their own deep-frozen forms.
(Animated, darling, was your tussock hair
When we lay shoreside naked there.
To my surprise not the lake water bluer
Than your blue eyes, nor drowsy noonday birds
As honey-liquid as your words.)

Our Lady Chapel secret cove
Was architraved blue above,
And green green the vestments
Of our invested love,
Love's altar a flat ridge of rock
Indifferent in the sun.

We laughed embraced in loneliness,
Two of us filling that private
Primeval Cathedral with a meaning.

Later returning loved loving
From the dark hush of undrowned bush
We found our altar congregated
With cabinet ministers
As many as it would take
Solemnly and without wonderment of praise
Pissing to raise
The level of the lake.

RANGI TANIRA HARRISON

Waikato te Awa

Katohia he wai mau,
Ka eke ki te puaha,
Ko Waikato te awa,
He piko he taniwha.

Kia tupato ra to hoe
kei tahuri koe,
I nga au kaha o
Waikato,
Whakamau to titiro
ki tawhiti,
Ko Taupiri te
maunga,
Ko Koroki te
tangata.

E hoe to waka ki
Ngaruawahia,
Turangawaewae,
Te kiingitanga,

Hoea to waka,
Ka u ki Kemureti,
Te Okohoroi o nga
tupuna

E hoe ana,
Ka tau ki Karapiro,
Titiro whakarunga
to kanohi,
Ki te tihi o te
Ihingarangi.

Kaati koa to hoe,
Titiro whakatakau
to kanohi,
Ko Maungatautari,
Ko Ngati Koroki,
Ko Arapuni te rohe
o te tuna e.

E piki ra to waka,
Ko Waipapa,
Maraetai,
Whakamaru,
Titiraupenga, he
maunga manu,
Ko Ngati Raukawa
e hoa e.

E tere to waka, ko
Pohaturoa,
Titiro kau atu ki te
tihi,
He parekura i hora,
I nga wa o mua ra.

Whaia te arawai a to
tupuna a Tia,
Naana i titi haere he
pou i muri i a ia,
Ko Atiamuri.

Kia ata haere atu ra ki
Ohakuri,
Te tomokanga atu,
Ki Orakei Korako,
Te whenua Waiariki,
Rua pehu pehu e.

E to i to waka,
I nga ara tiatia a Tia,
Tutuki ana ki te taheke hukahuka,
I tahuri ai to tupuna,
A Tamateapokaiwhenua e.

Tihei mauri ora, ka puta ka ora,
Tiu ana mei he manu rere rangi,
Ki roto ki nga wai marino,
O Taupo-nui-a-Tia e.
(Kua oti.)

Waikato is the River

Caught by the drag of the current at the river mouth. It is Waikato the river, a bend, a water guardian.
Paddle carefully lest you be overturned by the strong rapids of Waikato.
Focus your attention to the distance, Taupiri the mountain. Korokī the person.
Paddle your canoe to Ngaruawahia, Turangawaewae, the Kiingitanga.
Paddling your canoe to arrive at Kemureti Cambridge the washbowl of ancestors.
Paddle onwards, alight at Karapiro, raise your face to the peak of Te Ihingarangi.
Cease your paddling, turn your face to the right, it is Maungatautari, it is the tribe of Korokī.
It is Arapuni the district of the eel.
Your canoe ascends, it is Waipapa, Maraetai, Whakamaru. Titiraupenga, a bird sanctuary, it is the tribe of Raukawa, a friend.
Drift your canoe to Pohaturoa, look across to the peak, long ago there was a calamity.
Follow the waterway of your ancestor Tia, it was he who pegged out the poles behind him, it is Atiamuri.
Go carefully to Ohakuri, the entrance to Orakei Korako, the place of hot springs, exploding caves.
Drag your canoe over the rapids of Tia, crashing with the foaming falls, that overturned your ancestor Tamateapokaiwhenua,
The breath of life emerges alive. Soaring like a bird in flight, to the calm waters of Tauponui a Tia.
'Tis complete.

(translation by Herewini Easton)

DINAH HAWKEN

Hope

It is to do with trees:
being   amongst trees.

It is to do with tree-ferns:
mamaku, ponga, wheki.
Shelter under here
is so easily
understood.

You can see that trees
know how it is
to be bound
into the earth
and how it is to rise defiantly
into the sky.

It is to do with death:
the great slip in the valley:
when there is nothing left
but to postpone all travel
and wait
in the low gut of the gully
for water, wind and seeds.

It is to do with waiting.
Shall we wait with trees,
shall we wait with,
for, and under trees
since of all creatures
they know the most
about waiting, and waiting
and slowly strengthening,

is the great thing
in grief, we can do?

It is always bleak
at the beginning
but trees are calm
about nothing
which they believe
will give rise to something
flickering and swaying
as they are: so lucid
is their knowledge of green.

NOLEEN HOOD

Blaketown Beach

Bottles and cans
polystyrene pieces
string ... tyres
and broken glass
netting shredded
and embedded
in the sand
a very dead
seagull
beak strapped
with plastic tape
we couldn't pass
by ... we bent
in anguish
at the starving bones
and skin ... the flesh
thwarted ...
decomposed

Gathered together
in funeral pyre
the heap grew
as the beach wound
exposed
lay weeping

Plastic rubbish bags
filled with neglect
laziness
uncaring lolled
mouths gaping
as the accusing screech
of the dead gull
pierced the sky

PETER HOOPER

Three Pines by the Hohonu: A celebration

Late afternoon
as I turn down the lane to the bridge
and the burnt-umber sun musters clouds
sallow as tangled wool
on the half-wild sheep by the roadside.
I lean on the weathered railing and absorb
the repose of water
the unhurried pace of Hohonu,
its wrinkled evasive slide over sand and stone.

I follow my thighs and the calves of my legs,
my ankles and feet in the rhythm of loping
across stones, of balance, of sure-footedness.
Three tall pines on the hillside
deep-rooted in forgotten centuries
command the pool.
                                      I shuffle off old shoes,
socks, shrug out of my shirt, unzipp my jeans,
let all fall onto the log; they keep
a semblance of my shape, a kind of home,
soon to be inhabited again.
Naked upon the log on the bank,
under the thousand tremblings
of the trees on the hillside, I deeply
inhale, exhale, the smell of my
flesh is pleasant to me, the perfumed
air delights me, the water smooth
and brown as the body of my lover
invites me. I plunge to its embrace.
O! the shock of its seizure, the
sport of resistance, the gentle strength
as it holds me, bearing my braving.

I know I am clean and loved,
I am companioned, I dive to the stones,
I float, I wade and wallow.
I am loosened, dissolved; delivered from time,
I float in endless being, merge
fulfilled in the body of the world.
                         The sheep's-wool clouds
dye saffron before the sun. Emerging,
I rub myself quickly down, my flesh
smells of the river, I hurriedly pull on
shirt and jeans, the nipping sandflies
do not get much of me.
                      I amble back to
the road, pulling a stalk of grass,
gather a rata log to adorn
my hearth.
          I am become a part
of earth, of the river that flows forever,
I share the endurance of lost centuries
in the roots of the pines.

KERI HULME

Te Rapa, Te Tuhi, me Te Uira (or Playing with Fire)

      —do not be surprised at our anger
         the heartdeep anger of we who live here
         who have our families and futures here
         who revere
         the ocean of destiny
         you fart in—

We have not put our hope
our children's hope
in gunbearing ships:
we wait
and we talk together,
we sing and
we wait
while your military arrogance tests our patience—

the flotilla shudders
we shudder
as Papatūānuku shudders under her mantle of ocean
—fire and sea do not mix

    We are sea-people
    come from islands the world over
    sea-people are not intrinsically gentle
    storms and waves have seen to that
    yet we try to be pacific
    communally minded,
    caring for each other.

        you foul our home
        the very wind & water sicken

and our quiet seemingly helpless rage grows

hoon is the word here for the noisy and stupid
dangerous delinquent: that's you, hoons of the Pacific
There is no glory in what you do and
we do not tolerate hoons in our communities

Yes, we are small people from little islands
but we have sailed our hopes on the age-old path, that
   wideness of life,
our birth and making zone, Te Moananui a Kiwa—

                silent now upon the water the vaka is
                another witness to your depth-knell

                anciently, our waka-heke would halt
                watch in awe the wheels of fire
                underwater lightning
                we know there are older & wilder forces
                than your nasty flatulence—

you do not know
what you do

Listen: the wind is bringing curses
from mothers
of jellyfish babies;
listen still; the wind is bringing curses
from the old who nurse
their adult children, emaciated children,
cancerous children, who once sang
as they worked for you: listen again,
for the wind is bringing curses
not applause.

Small people, little islands—
we know that words live
long after the speakers die
and a tongue spear hits the heart
no matter how you guard against it.

                    Know we are loading our words,
                    fingering their sharpness,
                    readying them for unassailable flight:
                    we are breathing fire
                    into all our songs

so, do not be surprised
at our anger, and do not be surprised
when we dance in wild rejoicing
each time you play hoon, for we know the ending—

—do not be surprised
as your hearts begin to burn—

SAM HUNT

A Bottle Creek Blues

The wind can't blow any harder
the air's as heavy as Hell ...

I watched blue diesel smoke like mist
hanging on a high suburban hill:
wind I thought would blow it away
but the wind itself is diesel.

And yet the smoke disappeared
absorbed by that suburban hill:
the problem of disposal was
solved by the lungs of the people.

Two years ago we used to row
to an island here called Cockleshell:
gather cockles in a sack,
warm them up and gorge ourselves.

A friend I used to do this with
near died from typhoid fever:
they had the cockles analysed—
shit from down the coastline further.

Barefeet on the beach is madness,
this beach that was once made of sand:
the sun shines bright on broken glass,
cockles from Cockleshell Island are banned.

Sad protest songs are sung and heard
like this one here. And afterwards
the audience goes home convinced
the shit's cleared clean away with words ...

The wind can't blow any harder,
the air's too heavy for the birds.

DONALD McDONALD

Beginning Again

Someone felled the forest,
Someone fired the burn,
Someone scattered seed to grow,
Somebody in his turn
Watched the slow invasion
Of ragwort and fern.

I cut the fern and ragwort,
I burned the rotting log,
I dug out the ancient roots,
And drained the sodden bog,
And I burned all the rubbish
Where the plough might clog.

Now my team is waiting,
I toss aside my coat,
I tell the team to 'Get along'
And hiss the starting note,
And deep bite skeith and ploughshare
Into the earth's harsh throat.

1.9.39
First the axe and sawblade
Won then lost their gain,
Now the skeith and ploughshare
Cut where trees have lain.
Tell me, tell me Master,
Is their cutting vain?

Or are lads to follow
Where I came and went,
The file of those who struggle

To make the Earth relent;
The great, the Unsung Army,
Who knew what battle meant.

Not with axe and sawblade—
Not with skeith nor share—
What shall be the danger
That men must learn to dare?
Show them Lord, oh show them,
Speak and make them care.

CILLA McQUEEN

The Mess We Made at Port Chalmers

Tongue-stump of headland bandaged with concrete,
Obliterated beaches stacked with chopsticks.

All of this takes place in shallow time.

In deep time, the trees have already recovered the hills
and the machines rust, immobile, flaking away.
Healing, the land has shifted in its sleep.

All we would see if we were here
is seed-pods moving on the water.

HIRINI MELBOURNE

Aramoana

E kai nei te kino	The pain gnaws
I ahau nei	within me
E noho nei ki runga	As I sit on top of
O Ōtākou e	Otago peninsula
Kia whakatapuhia a Aramoana	Sanctify Aramoana
Raparapa ake nei e	The calm is
I te marino	disturbed
Ripiripi ake te tai	The tide of Huikaau
O Huikaau	in turbulence
Kia whakatapuhia a Aramoana	Sanctify Aramoana
Kei konei au	Here I am
Kei runga o Hautai	On top of Hautai
O Muaupoko	and Muaupoko
E kairangi nei	perplexed
Kei raro ra	Below my love
Ko te aroha	descends
E miri noa	to mingle with those
I te hunga moe te pō	who have passed
Kia whakatapuhia a Aramoana	Sanctify Aramoana
Toroa i te rangi	Albatross of the sky
Kaua rā e tau	Do not land
Kāore ō tāua tunga	There is no place for us
i tēnei ao	in this world
Kia whakatapuhia a Aramoana	Sanctify Aramoana

TRIXIE TE ARAMA MENZIES

Papakainga

A beach where earth runs out to sheltered channel
A delta formed when river changed its course
Fire in a ring of stones above which hung
Meat between two forked sticks, salt sky for savour—
That was a home, that distant campfire burning
With stolen fire, by our own hands plundered
Daring the gods, their glory for a hearth

Meat from that fire has lodged between my teeth
Bones from that meat lie buried in the sand
Shells from the pipis that the children gathered
Still rest upon that beach, heaped in the midden

Race memories of ancestral banquets haunt
That stony beach, that cooking place in earth
I rub rough sand into my winter skin
The dark blood drips upon the ground again
The scattered spirits gather close, reform
Cold ashes flicker into ghostly flame
And dancing shadows beckon, call my name

BARRY MITCALFE

The Road

The road brings trade
ah yes, a woman for a bottle
and a bead.

The road means progress.
Today I met progress
red-eyed

Howling down the road
his women gone
his children dead.

The road gives more than it takes
ah yes, tell that
to the man

Fighting for the land
or the gull on
the rock. When

The tide comes in
the gull
is gone.

The Maori road
took the way
of the land

The pakeha road
took the land
away.

SAANA MURRAY

Te Kōkota o Pārengarenga

Te kōkota o Pārengarenga
e tere nei ki ngā tai
e rere nei me he roimata.
Nā tauiwi ka ngaro e
ngā onepū, ngā taonga tūpuna—
kei hea rā he piringa
mō ngā hihi o te rā?

Te kōkota o Pārengarenga
te kainga o te pīngao,
o te huawai, o te kuaka.
Maranga mai e te rangatahi e,
pupuritia tō koutou mana
kei ngaro noa te tauranga waka e,
te onepu kōkota o ngā tūpuna
o Pārengarenga, o Ngāti Kurī.

The White Sands of Pārengarenga

The white sands of Pārengarenga
drift with the tides,
flow like tears.
The dunes, the ancestral treasures,
have been lost to strangers.
Where then will there be a refuge
for the dancing rays of the sun?

The white sands of Pārengarenga,
home of the pīngao,
of shellfish, of the godwit.

Arise, the younger generation,
take hold of your heritage
lest the canoe's landing place,
the white sands of the ancestors
of Pārengarenga, of Ngāti Kurī,
disappear forever.

JOHN NEWTON

Eldorado Poem

This is high summer, this is the river,
this is the shattering
impact of light on colour, on oats in the next paddock
blue as a mountain. The yellow, pollen-dusty tongues
that hang from the white furled cones of the arum lilies
echo the buttercups.
Duckweed, grey and crimson,
mirrors the red of the woolshed, the slow
rot in the yards.

The bottom bridge rattles underfoot, revealing
the river. A gap in the planking frames
twelve feet of water, a crayfish
three inches long
trailing a hard spiny
shadow on the blue mud,

frames a broken brick
worn orange and smooth by the weight
of freezing spring water
so clear it can only be seen in the movement of weed
or in the scoured ripple of the fleece
of a drowned sheep
rocking on the edge of the current.

The lilies shake down their golden
grain on the bulging
glass-bright skin of the river.
The big fish keeps to the bottom.
Bronze and purple, spade for a tail, fat back
scattered with ink-black glittering
spots, it tumbles

in the heavy current,
lunging after bullies,
rooting snails out of the stones.

Draw a line   with your eye
from the waving fleece
to the cluster of echoing lilies:

it's on that line you will find
the trout, the eye will single
it out when its
white mouth   opens.

CHRIS ORSMAN

Ornamental Gorse

It's ornamental where it's been
self-sown across the hogback,

obsequious and buttery,
cocking a snook at scars,

yellowing our quaint history
of occupation and reprise.

The spiny tangential crotch,
gullied and decorative,

I love from a distance,
a panorama over water

from lakeside to peninsula
where it's delicate in hollows,

or a topiary under heavens
cropped by the south wind.

I offer this crown of thorns,
for the pity of my countrymen

unconvinced of the beauty
of their reluctant emblem: this

burnt, hacked, blitzed
exotic.

EVELYN PATUAWA-NATHAN

Waikato Lament

Green wandering fingers
of kikuyu
prying into an old *kumara* pit
playing over limestone belly
and naked rock
have not quite covered,
cannot hide,
the faded emblems
of a land lost people.

Blood soaked, in time's
memory,
spirits of Taupiri
raise keening voices
anthem of injustice
echoing down
through the night.

MERIMERI PENFOLD

Tāmaki-makau-rau

Poua ki te hauāuru, hora ana ko Manukau.
Poua ki te rāwhiti, takoto ana Waitematā.
Maiangi ana i waenga, ko Tāmaki-makau-rau,
Tara pounamu o nehe rā, tau tuku iho ngā tūpuna—
Rangi e tū iho nei, Papanuku rā e tau nei!

Tāmaki e! Panuku e! Makau-rau e! Paneke e!
Tau whakairo a te wā, ngā tai e tangi nei—
Tematā, te hīnga o te rā, Manukau, te tōnga o te rā—
Takoto rā, te takoto roa, tōia mai nei e te wā
Mai i te heunga o te pō i te ao, ka ao, te ao hōu!

Tau wahangū o mata e, huri ake rā, ka tūohu.
Nei ngā whare kōrero: Maungawhau, Maungarei,
Maungakiekie, Rangitoto, he piringa no te tini,
He tohenga na te mano—tū atu ana he pakanga,
Hinga mai ana he parekura! A! Mau ana te wehiwehi!

Tau-iwi, maranga, e tau e—me he kāhui kūaka, auē,
Ki runga te tauranga nei! Ko Tāmaki-makau-rau!
Ka ū ki Waitematā, ka piki ki Maungawhau—
Ki Manukau, kua tau! Ki Maungarei, kua eke!
Hinga ana te wao a Tāne, tū mai ana ngā marae a Hōu!

Tamaki of a Hundred Lovers

Placed in the west, Manukau spreads out.
Placed in the east, Waitemata stretches out.
Between these rises Tamaki-makau-rau,
Greenstone pendant of the ages, the beloved passed down
From Rangi standing above, Papanuku lying here!

Tamaki shifts, Makau-rau changes!
The beloved of time and the sounding tides,
Temata at the rising sun, Manukau at the setting sun,
Lying there, lying long, hauled up by time
When night was parted from day, day, the new day!

Silent beloved of the ages, turn and see!
Here are the homes of speech: Maungawhau, Maungarei,
Maungakiekie, Rangitoto, grasped by many,
Contested by multitudes—battles fought,
And battles lost! Oh what terror!

Strangers, come, settle like godwits
On the landing-place! It is Tamaki-makau-rau!
They land at Waitemata, climb Maungawhau,
They alight at Manukau, mount Maungarei!
The forest of Tane falls, the marae of the new men lie here!

(translation by Margaret Orbell)

VERNICE WINEERA PERE

Song from Kapiti

Some people there are
who survive
on the promise of summer.
Such are the inhabitants
of the Paekakariki coast,
in dwellings nestled against
cliff-face and ragged clay
in the full winter fury
of the open ocean.
I watch a lone sea-bird
who sits woodenly
on a wind-ravaged crag,
his feathers ruffled
by the cold southerly
straight from the Antarctic.
The sparse toe-toe
lean with the wind,
and trail feathered fronds
over the wild, grey beach.
They reach chill fingers
into my heart.
I am that bird
frozen by the southern wind,
my wings wooden
in the cold salt air.
I am the child of the Ngati-Toa,
seeking my place
in a mainland society.
I am she learning to sing
the sad-sweet songs of a people's soul.
I am the lone bird
alive in a limbo of longing,

enduring the winter world,
surviving
on the slim promise
of a future summer.

ROMA POTIKI

A Cloak and Taiaha Journey

Dark shapes sing in my head
on the way to the Cape
a cloak and taiaha journey,
dusty road
a narrow skimp of dry land
on the way to an airless dive
—leap to infinity.

I nearly fall asleep
and in the near dreaming
see the skinny arms and bodies
of old ones
etched, chiselled and curled
their unearthly features holding the tatau of land,
sky and water.

They do not speak in the language of sound
but directly to the ones seeing them.
The black whorls that carve their limbs
are strong in shape
and delicately meet on each ridge,
each peak and feature.
When they dance it is light and deep
their thudding hardly noticed till you realise
a pattern of breathing.

I do not remember being born
and these ones similarly placed
continue to raise skin,
blow soft stinging currents on hair.

If they had stones in their hands
you'd feel them on the back of your neck
before you fell.

They are here in waves
in melting, creosote washes
the speechless tupuna stamping out messages—
old codes
drop by drop, oil on the tongue

seeds.

They are everywhere
in the hills and mounds of this place
the women and men, the others.

Though I see a few there are many,
sinking and rising they renew themselves—
communicate.

HARRY RICKETTS

Memo for Horace

You're not going to believe this
but they're now imagining the probable effects
of a 'nuclear winter'.

No idea what I'm talking about?
There isn't time for a lesson in C20 science
but it's the latest explanation

and explains incidentally
why I sit here on this particular Spring evening
reading 'Exegi monumentum aere perennius'.

In this new hypothetical world, you see,
it will not matter what monuments
are 'more durable than bronze'

or 'higher than Pharaoh's pyramids'.
No part of us will survive for long
the 'angry wind' and 'hungry rain'
of that last incandescent translation.

KEITH SINCLAIR

The Bomb is Made

The bomb is made will drop on Rangitoto.
Be kind to one another, kiss a little
And let love-making imperceptibly
Grow inwards from a kiss. I've done with soldiering,
Though every day my leave-pass may expire.

The bomb is made will drop on Rangitoto.
The cell of death is formed that multiplied
Will occupy the lung, exclude the air.
Be kind to one another, kiss a little—
The first goodbye might one day last forever.

The bomb is made will drop on Rangitoto.
The hand is born that gropes to press the button.
The prodigal grey generals conspire
To dissipate the birth-right of the Asians.
Be kind to one another, kiss a little.

The bomb is made will drop on Rangitoto.
The plane that takes off persons in a hurry
Is only metaphorically leaving town,
So if we linger we will be on time.
Be kind to one another, kiss a little.

The bomb is made will drop on Rangitoto.
I do not want to see that sun-burned harbour,
Islandless as moon, red-skied again,
Its tide unblossomed, sifting wastes of ash.
Be kind to one another, kiss a little,
Our only weapon is this gentleness.

J.C. STURM

As the Godwits Fly

for Frank

After days of coming and going
Taking off, circling, landing again
So lightly and exactly dropping
Into line upon line
Settling and resettling themselves
Always facing into the wind,
Hunched low on the rocks
Like runners on their marks
Poised ready for that final
Instinctive command to
Go!
The godwits have gone
Every one, all together
Taking summer with them
To fly a private longitude
From the bottom to the top
Of the world
Where northern winters
Thaw to sudden springs.

I may never see them again.
Certainly some will fall
Down shafts of turbulence,
Or simply tumble out of
A blue and cloudless sky
Small grey and white feather bundles
Still warm,
Or falter and flutter sideways
Off course
Down a different longitude

To make a final, unplanned landfall,
While far above them
Sister hearts and brother wings
Keep on beating, beating
Without pause
Northward into the wind.

Right now I would give
Anything
To be that child again
Watching in wonder
For the first time
A cloud of white birds rising
And heading north;
To live again from then
To now exactly as it was
With no correction or variation,
Living it straight
As the godwits fly
My longitude from end to end
And keep on flying, flying
Without pause
Past my final winter
Thawing to its spring.

APIRANA TAYLOR

Feelings and Memories of a Kuia

These Māori today
are not Māori any more
I don't know what they are

I remember the old people
they were polite
and they liked to talk
they walked at a leisurely pace
but always got things done

Today my mokopuna always rush
yet never seem to do a thing
they hardly say a word to me
and they don't look happy

When I was a girl
we had big gardens
all us children worked in them
and the kuia and koro
used to make jokes
they cooked us a big kai in the morning
we were very happy

Sometimes
we went to the lake
we'd get some fern
and tie it into bundles
with akeake
and toss the fern
into the water

When we pulled up the bundles
we'd shake them
and lots of fish
would fall out

In the bush
there were plenty
of fat juicy pigeon
the river
was full of eels
dinner swam past
all we had to do
was catch it

The bush
got chopped down
the river dammed
and the lake polluted

It seems to me
the closer
the Pākehā got to us
the more difficult
he made it for us to live

Then came the war
I think
a lot of Māori men
who would have been
great leaders
of our people
got killed overseas
and for what

So we could live in quarter-acre sections
when once we had
more land
than my eye could see

So we could eat
hamburgers
when once we got
fresh kai

So in the end
we could be lost
and unhappy
and not know why

NGAHUIA TE AWEKOTUKU

Mururoa / Moruroa

Murua o matou hara
Forgive us our trespasses
Murua ... murua
Forgive us, Mururoa
Wipe out these wrongs, forgive
Mururoa
Long distance forgiving

So the saltwarm of your
waters move
ancient in our blood
remembering
canoes seeded cast
a thousand years ago
from homeland islands
salt warm
remembering

And now
hard basalt heaving
liquefies: seething
grey foam
polite so pleased
the french clap

before Pape'ete burns

Murua o matou hara: forgive
Do foreign powers with
their foreign god
deserve forgiveness?
Mururoa Mururoa

long distance forgiving
of us: remembering
protecting
homeland islands
homeland islands
salt warm
remembering

We sent you boats.
We tried, though we
weren't on them.
Murua, Murua,
Mururoa.

(Mururoa, in Māori, could translate as 'long distance forgiving'. The first line of this poem is lifted from the Māori version of The Lord's Prayer.)

DENYS TRUSSELL

Ore: An ecological poem

All they do is savage and violent in accordance with Nature around them; and this exercise of their passions they mistake for progress. In all things they exhibit immense energy, and herein alone they show they are subject to the titanic and terrible presence around them. **(D'Arcy Cresswell, 1939)**

One wave of matter breaking
from oceanic depths.

Now a reef, man-corroded;
a lode shattered
by a storm of engines.

It is a land of ores,
broken anchor to its people.

Our history here,
the leached alluvial golds;
our meaning here,
this haggard ground.
The hill scarred tall relic
of our mineral famine.
The earth-deep seam
of anthracite hollowed.

Lost to the secret
places of ourselves,
we map the enormous
wage of iron that sleeps
along the coasts.

What is human
in us breaks
so easily against
our stone indifference.
What is human
in us drowns
by depth of love
or hate / darkens
like lustre in the blood.

What is insensate
in us continues
its blind search
of intricate stone
to forget the voided self.
What is insensate
in us extracts
toxic riches from the rock
to make of matter's magnetic
web our human burial.

It is the land of its people,
their grave of ores.

BRIAN TURNER

Van Morrison in Central Otago

Take me back to the days
before rock and roll,
Life with Dexter and
the Goons, days when
Dad and Dave still bumbled
in Snake Gully
and the skies were not
cloudy all day, before

I saw tussock, heard it
speaking in tongues
and chanting with the westerly:
What's productive here
is what's in your heart,
sworn through your eyes,
ears, the flitter of the
wind in your hair;

the smell, the taste
of air from the mountains,
off flats where the river
runs from somewhere north
to somewhere south and the sky's
forever. It's not picturesque,
it's essential, almost
grand, and it aches

like the rhythms of truth
scornful of tittle-tattle.
You have to be here, you
have to feel the deep
slow surge of the hills,

the cloak of before, the wrench
of beyond. You know
what, you know
not. And that's what
makes it heart-stopping,
articulate, hurtful
like resuscitation.
You cannot bear to use
the word again again
when driven by an urge
to begin to begin.

HONE TUWHARE

No Ordinary Sun

Tree let your arms fall:
raise them not sharply in supplication
to the bright enhaloed cloud.
Let your arms lack toughness and
resilience for this is no mere axe
to blunt, nor fire to smother.

Your sap shall not rise again
to the moon's pull.
No more incline a deferential head
to the wind's talk, or stir
to the tickle of coursing rain.

Your former shagginess shall not be
wreathed with the delightful flight
of birds nor shield
nor cool the ardour of unheeding
lovers from the monstrous sun.

Tree let your naked arms fall
nor extend vain entreaties to the radiant ball.
This is no gallant monsoon's flash,
no dashing trade wind's blast.
The fading green of your magic
emanations shall not make pure again
these polluted skies ... for this
is no ordinary sun.

O tree
in the shadowless mountains
the white plains and
the drab sea floor
your end at last is written.

HONE TUWHARE

Papa-tu-a-Nuku (Earth Mother)

We are stroking, caressing the spine
of the land.

We are massaging the ricked
back of the land

With our sore but ever-loving feet.
Hell, she loves it!

Squirming, the land wriggles
in delight.

We love her.

MURU WALTERS

Haka: He huruhuru toroa

He huruhuru toroa te rere o te kapua
      ka pae ki runga o te tihi o Pukekura e
Aūe! I ahu mai koe i Hawaiki nui
      Hawaiki roa, Hawaiki pāmamao e
Aūe! Nō te ao pani, nō te ao mahue
      kua hoki mai koe ki te awhi
      i tō whenua
E! I roto i ngā whenua raupatu
Noho i te whakamā! Noho i te momori!
Noho i te whakamā! Noho i te momori!
Hore kau he oranga kei roto i
      tōku ngākau
Hōmai he whakahauora!
Toroa nui! Toroa roa!
Toroa rere mai! Toroa pāmamao e!
Auē! He tau pōuriuri
      He tau pōtangotango
      He tau pō tiwhatiwha e!
Hena! Te marama o runga o
      Aramoana tiaho ki te kāinga e
Auē! Pōuri nui! Pōuri roa!
Kia kite i a tauiwi nui!
Tauiwi roa! Tauiwi pāmamao e!
Auē! Whakaparu ana i tō
      tātou whenua
Hena! Mātakitaki! Mātakitaki!
      Mātakitaki! Mātakitaki!
Ki ngā kaipuke o tauiwi
      E rere mai nei!
      E rere mai nei!
Kei hea ngā waka Māori?
Kua tangohia e!

Kua tangohia e!
Ki te kōpū o te tai
I a Kōpūtai e!
I a Kōpūtai e!
Maringi ana ngā roimata ki raro
Ka kake anō ki runga
Tīheia ngā tini mahara
Ka awhi i roto i te pōuri
Te poupou whakairo ko Āraiteuru
Hei kāmaka!
Mō te kāinga e! Hi!

Haka: The feathered albatross

Like a cloud the feathered albatross
    alights on the summit of Pukekura
Aue! You came from the great Hawaiki
    the lofty Hawaiki, the distant Hawaiki
Aue! From a world of orphans, from a world deserted
    you have returned to embrace
    your homeland
To your confiscated homeland
Remaining in shame! Remaining bare!
My heart is sick within me
Inspire me!
Great albatross! Lofty albatross!
Flying albatross! Distant albatross!
Aue! A time of dark night
    A time of intense darkness
    A time of intense sadness!
Look! The moon over Aramoana
    illuminates the ancestral home
Aue! Great darkness! Deep darkness!
Witnessing great foreigners!

Lofty foreigners! Distant foreigners!
Aue! Polluting
        our land
Look! Look! Look!
        Look! Look!
At the foreign ships
        flying here
        flying here
But where are the Maori canoes?
Taken!
They have been taken
In to the stomach of the tide
By Koputai!
By Koputai!
Tears fall underground
And rise again
With a great burden of memories
Embraced in the darkness
Is the carved post Araiteuru
The foundation
Of the ancestral home!

(Translation by Muru Walters)

IAN WEDDE

from **Pathway to the Sea**

to A.R. Ammons

      {...} yeah, I heard
          they wanted to build an

ALUMINIUM
    SMELTER
        at
Aramoana, the sea-gate, & someone's bound to direct
    more effort that
        way soon: listen, there's
birds out there, we're
    back with those lovers, the buoyancy
        & updraught of some kind of

mutual understanding of what
    service is, of the fact that
        a thing being easy doesn't
make it available or passive:
    listen, effort's got to be right
        directed, that's
all, the catering's amazing, everything
    proceeds, citizen, sometimes
        it's hard work, but you're

engaged, you want
    to keep practical & ideal
        together, you're
for life, you know that happiness
    has to do with yes
        drains & that nature

    like a pear tree
        must be served before
            it'll serve you, you

don't want your children's
    children paying
        your blood-
money, citizen, you're
    for a different sort
        of continuity, you want
to live the way
    you want
        to, you want to keep

your structures up, you
    want elevation,
        you're ready to do
your share, you'll dig your field-
    drain & you'll
        keep your shit out
of the water supply:
    you want to
        serve & to be left alone

to serve & be served,
    understanding tough
        materials, marl & old timber,
the rich claggy rind
    of the world where
        dinosaurs once
were kings: well they're gone now though
    they survived longer
        than we have

           yet, but then we know, don't we,
                citizen, that there's nowhere
                      to defect to, & that
living in the
       universe doesn't
              leave you
any place to chuck
        stuff off
              of.

ALBERT WENDT

In the Midnight Ocean of his Sleep

In the midnight ocean
of his sleep an impatient moon
fishes for the morning,
its phosphorescent hooks flash
like glow-worms on
the hides of leaping dolphins
barking at the dark.

No clouds, no whisper of rain.
Silver mice nestle in the warmth
of his head. Where mountains crouch
is the dizzy scent of mosooi.
The longtailed tropic bird circles
and circles sleekly the rims
of his eyes in a slow eternity.

Alive with squealing flying-foxes
furred soft like his breath,
a banyan fingers down into
the black earth to oil his ancestors' bones.
He watches and believes he can
resurrect his dead with one
circling song of his hands.

In that song his mother sits
combing her long black hair.
Terror is the succulent mango fruit
he plucks and devours,
the mango stone he plants in her shadow.
His world is without fear.
I am not in his dream.

When he wakes he'll walk on
the water to islands I can't reach—
moon, morning, banyan, flying-foxes
and his ancestors snared in
the Maui sinews of his hair,
the dazzling tropic bird his compass,
dolphins whispering the way.

Hold back the dawn,
cut the moon's fishhooks,
I must leap into my son's
ocean of dreaming so when
he wakes I too can go
with him to discover where
the wise dolphins die.

Now

Twenty-first-century ecopoetry

As mainstream conversations about Aotearoa New Zealand's history turn from a predominant Pākehā narrative of European discovery and settlement to acknowledgement of original habitation by Māori and the devastating effects of colonisation, ecopoetry reveals a corresponding shifting view of ecological connectedness and loss. Climate change, mining and ocean pollution are viewed in terms of their damaging effects on local ecologies and cultures throughout the Pacific region. Te Moana-nui-a-Kiwa, the Pacific Ocean, is 'a food basket' in Te Kahu Rolleston's poem 'The Rena' (2014). In Karlo Mila's 2020 poetry collection, the ocean is 'goddess muscle'. Pacific poets Tusiata Avia, Selina Tusitala Marsh and Karlo Mila, and Māori poets Vaughan Rapatahana and Robert Sullivan write of cultural connection between this country's ecologies and the Pacific region stretching back centuries. These connections tie Aotearoa New Zealand to its location within the Pacific not only through geography but also through culture.

In the twenty-first century, colonisation continues to be evoked as an ecological and cultural intrusion. In 'my whenua' (2004), Apirana Taylor writes, 'my whenua lies buried at the Karori dump / beneath asbestos dirty nappies and waste' and 'there are oxidation ponds / where once we caught the kokopu'. Hone Tuwhare, in his posthumously published 'Pupurangi (Kauri Snail shell)' (2011), describes 'this sad environ / of depletion'. Pākehā poets, too, connect colonisation and ecological degradation. Introduced species are represented as ecological destroyers: in 'The Land without Teeth' (2019), Rebecca Hawkes calls gorse (and her poet speaker) an 'impregnable alien … your favourite coloniser … me & my entire invasive species / consuming the landscape once so toothless'; Gail Ingram's ironic 'Recipe for a Unitary State' (2017) lists ingredients that include stoat, rabbit, cows and calicivirus; and Jonathan Cweorth's 'Ghost Stoat' (2107) 'set[s] the tui gagging'. Mining is ecological degradation in Leicester Kyle's *An Aside:* Advice to the Rehabilitator' (2004), which lists indigenous trees, tussock, birds, geckos and snails that would be dislocated by the reopening of the open cast coal mine at Millerton on the west coast of the South Island.

There is rejoicing in some aspects of nature – Lynn Davidson an otter, Fiona Farrell an eel, Sue Wootton icebergs, Briar Wood whai – but such moments of admiration are often undercut with the spectre of loss. 'Every year they're moving nearer—', writes Wood in 'Whai' (2017), 'this bay's being fished out.' In 'A Behoovement' (2017), Wootton asks, 'What is the soul to do, if the icebergs melt?'

While twenty-first-century ecopoems are often bleak, at times, there is hope. Carolyn McCurdie's 'Ends' (2017) suggests the power of unity within the counterculture to effect change:

> song coming.
> Song coming, beyond any
> we've ever before lifted in song.

And on a 'pre-cyclonic post- / apocalypse-that-never-came rainy early- / summer Christmas Eve', Kiri Piahana-Wong finds strength in sky father and earth mother:

> —the sky's water: all the wondrous light
> weeping joyous tears of the sky god,
> Ranginui, running down my side and
> into the earth, Papatūānuku, and then
>
> settling there.

Even with the passing of time, regret over indigenous birds lost during the Victorian era remains in the twenty-first century. Seven poems in this section lament the loss of the huia, which became extinct in 1907, an absence keenly felt in the blank space of Sam Sampson's 'Erasure' (2014). The bird, with its distinctive white-tipped, black tail feathers and the long, curved beaks of the female, continues to haunt, not only because of its absence but also because of what remains: feathers 'kept in waka huia, / treasure boxes' (Anna Jackson, 'Huia', 2001), 'greed and anguish' (Bill Manhire, 'Huia', 2020), 'brooches, adorned with tiny chains' (Alison Glenny, 'Candle', 2021), 'lyric mimicry' (Emma Neale, 'Huia', 2015). In Neale's poem, the only recorded recreation of the huia's call is lost due to a technological mishap. Hinemoana Baker's 'Huia, 1950s' (2004) reminds us that the recorded imitation was whistled by 'the huia-trapper', revealing something of the complexity of ecological loss in Aotearoa New Zealand. 'I want to tug / something out of him', she writes. In 'The Anthropocene *circa 2016*' (2018) Helen Heath portrays the recording of Hēnare Hāmana whistling a huia call as 'a dead man calling a dead bird, an echo from a machine', suggesting the impossibility of keeping hold of something once it is lost.

Anthropocene, the term for the current geological age during which human activity has been the dominant influence on climate and the environment, enters this country's ecopoetry in the twenty-first century. Heath's poem explores the influence of technology on urban nature, such as the tūī, which mimics phone ringtones and other human sounds. Her poem references adaptions that are occurring here and now in Aotearoa New Zealand as human sounds distort nature, altering our comprehension of what nature is.

Ash Davida Jane goes one step further by conjuring a future vision in her poem '2050' in which the sky is deliberately polluted, and life is lived beneath overhead projected images, 'squares of blue / with white blurs passing across.' The 'false sky' is chilling enough, yet even more so when Jane suggests that memories of the real sky could fade completely. 'I'm afraid', she writes, 'I could forget / there was ever another way.' A sense of despair turns to hopelessness in Van Mei's 'There's Real Mānuka Honey in Heaven' (2020), which looks even further ahead to the year 3000. In a world without people, 'the tuatara sings a eulogy to the end of the anthropocene ... as the cockroaches wait for spring'. The effects of inaction due to climate change deniers are undeniable in 'All That Summer' (2016) by Tim Jones, which imagines a future when Wellington, at least, is afloat due to sea level rise, and in Sue Fitchett's 'Story Lines' (2017), which suggests ecological loss through the blank space of erasure. At times, even the global present is filled with a sense of dread. The white pages of Alison Glenny's *Bird Collector* (2021) shriek absence, as does the title of her 2018 collection of Antarctica poems, *The Farewell Tourist*.

At a local level, even Brian Turner's poems, which since 1978 expressed a sense of deep connection with and joy in the presence of Central Otago's remote rivers and ranges, assume an attitude of despair this century. In 'Lament for the Taieri River' (2008), the Romantic language Turner used in his earlier poems to evoke a sense of love for the region is replaced with harsh words and assertions that heighten a sense of disconnection between the poet speaker and the river:

> The Taieri River stinks from effluent,
> is soured by pesticides and fertilisers.
>
> The flow is sluggish, low;
> shit and muck slosh into the river
> where cattle mill and piss and break down banks

By shifting from the figurative language that characterised his earlier work to the prosaic language of ecopolemic, Turner substitutes a sense of communion with the river for a sense of distance. Fishing is described only through reminiscence and serves as a reminder of the river's unpolluted past. But it is not only trout and their rivers that Turner mourns. 'Lament for the Taieri River' comprehends the loss of those places that afforded a sense of solace from modernity.

In her collection *Flow: Whanganui River Poems* (2017), Airini Beautrais casts trout fishing as a symbol of connection between people and nature and evokes the risk of losing trout habitats due to human-caused pollution. She extends Turner and John Newton's twentieth-century portrayals of trout and fishing through a critique of the introduction of trout into Aotearoa New Zealand rivers by European settlers in the 1860s. 'Trout / *Oncorhynchus mykiss / Salmo trutta*' proposes that the practice of anglers coming from other countries to experience trout fishing in Aotearoa New Zealand entrenches ecological colonialism over less dominant, indigenous species:

> The small and the slimy are fit for ignoring
>   the roaring
>   of rapid calls to the rover.

The poem explicitly relates colonial dominance of the exotic over the indigenous, 'That which was here first is caught in the jawing / the gnawing', as 'newcomers spread and their numbers keep swelling.' By framing trout as ecological aggressors, Beautrais complicates those earlier Romantic visions of their surroundings as places of solace from modernity. Instead, she exposes trout as intruders and their habitats as places of colonial construction where indigenous fish became extinct.

Freshwater pollution is the subject of Jacq Carter's 'If I Am the River and the River is Me' (2013), which was composed for 'The River Talks', a call to action to support the rejuvenation of the Ōmaru River in Glen Innes, Auckland. For Carter, the cultural loss associated with the polluted river is both personal and communal. Her poem speaks not only to the loss of fish but also of Māoritanga. '[T]he paru has stolen my clarity', she writes, suggesting not only a reduction of water quality but a blurring of a sense of self. The river's name, Ōmaru, a place of shelter, has been polluted too by a loss of knowledge of the river's cultural past which fashioned its significance:

> If I am the river and the river is me
> then few of us actually know my history
> Ōmaru could mean absolutely anything
> making me just a dirty stream

In Rangi Faith's 'Losing our Mana' (2005), the loss of eels from a river represents not only a shortage of food, 'there / hasn't been an eel / caught / for the last three tangi', but also a forfeiture of mana. As the title illustrates and the poem repeats, 'there is no kai / in the river, / we are losing our mana.' The absence of eels exemplifies not only an ecological loss but a cultural loss as well. Such ramifications stemming from the usurpation of indigenous species by introduced species frame the nature constructed by colonialism as ecological and cultural violation.

The degrading effects of colonisation, which have been referenced in Aotearoa New Zealand ecopoetry since the nineteenth century, extend this century across Pacific national boundaries. Tusiata Avia, in her collection *The Savage Coloniser Book* (2020), employs anger and irony to expose the devastation of colonisation in the region. 'Fucking St Barbara (i)' frames the Simberi open pit gold mine in Papua New Guinea as rape:

> He prowls a mile under the surface of my skin
> He has large grinding wheels in his cock
> it rips me up like I'm concrete, like I'm the sea floor.

The poem asks about the effects of the mine on local people and imported workers, 'the 1100 indigenous people / the Cape Coloureds brought over from South Africa', suggesting global cultural impacts. It concludes that the island cannot escape the ravages of global capitalism. 'Where can I hide?' it asks. 'Even the sea is full of rapists.'

The cultural connectedness of the Pacific Ocean, Te Moana-nui-a-Kiwa, is iterated by Karlo Mila. 'We are a water world', she writes in 'Matariki: A Call to Kāinga' (2020), which takes stock of the polluted Pacific:

> Oh ocean, we ask too much of you.
> Gagged, plastic-bagged.
> You hold too much heat
> on our behalf.

Selina Tusitala Marsh recounts a sense of helplessness in the face of sea-level rise caused by climate change in 'Girl from Tuvalu':

> this week her name is Siligia
> next week her name will be
> Girl from Tuvalu: Environmental Refugee

In 'Old Bones' (2017), John Howell portrays the effects of sea-level rise on a Tuvaluan cemetery where 'graves are waterlogged canoes.' Vaughan Rapatahana's 'Topside, Nauru' (2013) depicts the ecological catastrophe on Nauru where strip mining of the phosphate plateau known as topside has left a wasteland of limestone outcrops uninhabitable by plants, birds or people:

> there is nothing else here
> frigate birds stay away,
> and even the leaves have left
>
> just jagged monoliths:
> dead men.

In Te Kahu Rolleston's 'The Rena' (2013), the 2011 grounding and subsequent break-up of the container ship on the Astrolabe Reef in the Bay of Plenty is represented as a breach of protected food sources: 'how dare you, / poison the swells and the realm of Tangaroa'. In this poem, the Pacific Ocean is 'a food basket and store, / payments made, with the practice of Kaitiakitanga, Tikanga, and L.O.R.E Law.' Delays in cleaning up the Rena's fuel and containers is 'a culture clash, / Money vs Mana.'

> The sea ... Well that's a *resource* to followers of capitalism,
> but what's a *re-source*, to our *Mauri-source*,
> my people's very essence of living.

The representation of the Pacific Ocean as a part of Indigenous 'essence of living' expands the effects of pollution, sea-level rise and exploitation of islands through mining, from physical contamination to cultural desecration of the region as well.

In 'The Land without Teeth' (2019), Rebecca Hawkes appears to mock those poets whose inheritance stems from British and American Romantics when she writes, 'I get with the dirt ...

>   but this body's no alpine native
> don't mistake me for some mystic nationalist
> in my whiteness   & my rhetoric of wilderness
>
> I am introduced to this place
> like the gravel track tattooed on the slopes
> & the unstrung barbed wire

Hawkes's speaker is a self-conscious coloniser, like gorse, an exotic weed that can't be removed, 'you couldn't unroot me if you tried.'

Dinah Hawken is another Pākehā poet who self-consciously explores colonial ecological and cultural violence. Her collection *There is no Harbour* (2019) traces the arrival of her great-great-grandparents, Jane and Joseph, to Taranaki from England in 1845. In 'Losing everything', she writes, 'I am the beneficiary of injustice … And of Jane and Joseph's labour / and their love.' Hawken's poem reflects on the hardships faced by early colonial settlers alongside the ruinous effects of Pākehā settlement on local Māori:

> While they 'did something for a family'
> there is no harbour
> —we are tainted by confiscation.

'There is no harbour' suggests there is no place of shelter or resolution for Pākehā in the face of 'the depth of injustice Māori have endured in Taranaki'. Her poems portray the trials of settler life alongside land confiscation and the consequent loss of a Māori lifestyle dependent upon indigenous ecologies, thereby linking her own ancestry with a contemporary sense of sorrow for colonial ecological and cultural loss.

Already this century, poets have produced an abundance of ecopoems that reflect on how we comprehend ecologies. Regret for human-caused ecological losses and associated cultural losses are continuing themes. Increasingly, concern about the future of local and Pacific-wide ecologies is evoked. Alongside issues that inspire ecopoets everywhere are comprehensions of ecological and cultural losses and senses of belonging specific to Aotearoa New Zealand. Understandings of nature and the human relationship with it from the perspectives of tangata whenua and settlers within a settler colonial context are at the forefront of our literature as ecopoetry continues to trace shifting attitudes to ecologies and to each other.

JENNY ARGANTE

Seal Mourning

So rushed the morning came
along the coast
I saw them bobbing, supple and rounded
with whiskered baby-faces
eyes moist and mute with messages
my guilt would not translate;
and through the waters floundered
on a high note calling
shrill notes that hurt me, listening
as others would not listen

And the sea flooded out
its detritus, and flooded in
and floated dead and dying
and the resolute survivors
haunting us with reproach
of round and startled eyes

The seals are mammals, giving birth like us.
They suckle infants who will not survive
and, in remission, frolic
among the waters that will sign
oblivion

My sons and daughters, quickly breed your children
lest I, old granny, dreaming of the past
draw pictures, tell the tale of 'once upon a time'
when seas were clean and offered sanctuary
and joyous creatures dived and dwelt within;
creatures like this, my children,
whales and dolphins plunging
and grey seals—o, long gone!—that with their antics
mimicked babes and lovers

All squandered now—long gone, my little ones,
those voices hushed and fluting into silence
that through kind oceans sang the sailors home.

BRIDGET AUCHMUTY

Marginalia

For a while we lived by the ocean
instead of by the clock. Tides came and went

lapped yards from our pillow or roared
way out reefing on Farewell Spit.

We chose, he and I, to live on the edge
of things: world, breadline, legality,

comfortable under the radar
of building permits and the IRD

as if we occupied one of those baches on the Boulder Bank
the in-between domain where the only constant

is perpetual change, rendered central
like the threadbare linen sheets my grandmother stitched

sides to middle to extend their life.
But the seam unsettles … and what I'm looking at

is marginalia, how those hand-jottings draw the eye
like flotsam: all that floats onto the margins

of the shore—plastic bags, polystyrene
kickboards fishing floats gill nets

pastimes become perilous—
washed up entangles

strangles
drowns.

TUSIATA AVIA

Fucking St Barbara (i)

What happens when you fall for an Australian goldmine?
I'm not speaking in metaphors here, I mean:
the mechanics, the wall, the maintenance
the stock exchange for the March 31 quarter
the whole St Barbara Company Ltd who own
the island economy on Simberi
the 1100 indigenous people
the Cape Coloureds brought over from South Africa
who look like The Rock but with 2 grams of gold per tonne of earth.

Corporate responsibilities stop in the hole
between my legs. You've got to dig a big hole
to get to the ore body
you might even have to remove a mountain
and then dig another big hole in the ground
the size of twenty-one football fields
to get to this ore body's ass.

How long can I keep his hands off?
How long?
Where can I hide?
Inside the water?
Even the sea is full of rapists.

He prowls a mile under the surface of my skin
He has large grinding wheels in his cock
it rips me up like I'm concrete, like I'm the sea floor.

He wins the St Barbara 2020 Digger of the Year award
for his performance: 93,000 Ounces of Gold out of Simberi.
He kisses me before his walk to the podium
as they play 'You Ain't Seen Nothing Yet'.

HINEMOANA BAKER

Huia, 1950s

the huia-trapper

whistles the song
I try to resist

I want to tug
something out of him

the radio voice says
believed to be extinct

AIRINI BEAUTRAIS

Trout / *Oncorhynchus mykiss* / *Salmo trutta*

He with his flexing
                ray-fins at the ready
she scrapes her redd
                in the bed
                              of fine gravel
where water will travel
                in riffle soft-peaking.
Now he moves, sexing
                the roe lying steady
spills his milt, spread
                to the head
                              at a level
to let it unravel
                and meet what it's seeking.

Yellow sac'd alevin clustered in hiding
colliding
        together, translucent, light-bending
freckled and blending
                to stone-colour, matching
each roll to the current's tug, latching, unlatching.

Bug-eyed and bulbous, their yolk-spots subsiding,
backsliding
        they break from their cracks to the wending
of current, fast-tending
                to gulping and snatching
grown in the flicker of seek-and-dispatching.

Fingerlings flashing
                in blurring reactions
swift in their shifting,

                      gulp-sifting
                                    and feeding
on scoops of fresh seeding
                                on small fry, on all things.

Rising and splashing
                        to grasp dark refractions
of drowned insects drifting
bob-lifting
                                    black-beading
the surface. A weeding
                          of weakness, of small things.

There was the water, as though it were waiting
creating
                an emptiness, ripe for the taking.
Sport in the making,
                    come fry in a swilling
of barrel by cartload. The rivers are filling.

Flies for the flicking, and hooks for the baiting
fixating
                in hunger, slow stares without breaking.
No easy slaking
                    this thirst, or this thrilling
the line gathers weight, and the buckets are spilling.

There is beauty
                    in marking a cluster
each to his stone
              wet to bone
                            in the pooling
where the wide-spooling
                    hook shimmers, dangling.

We have a duty
                to strengthen and muster
keep what's our own
                what we've grown,
                              overruling
anything spoiling
                the pleasures of angling.

That which was here first is caught in the jawing,
the gnawing
                of newness works to uncover.
Seek to recover
            then fall to a culling:
newcomers spread and their numbers keep swelling.

The small and the slimy are fit for ignoring
the roaring
                of rapid calls to the rover.
All the world over
                      tales in the telling
Keep travellers coming, each one of them calling,

Saying: Come, fish, break from your hollow
Be rose-moled, be of sufficient weight
To meet your fate
                on the thin line you follow.
Gulp, swallow
              this darting lure, its yearning.
Come to my hook, fish, and keep returning.

BRENDA BURKE

Feature Battle

In the
FEATURE BATTLE
we have

'sources of wet' versus **'sources of dry'**

'wet' won the toss and opens

'OK, then, big, big, oceans!

'the sun! our *friend*

rainfall in the mountains

burning. oil, coal,
gas and stuff

rivers, streams, creeks

thirsty super-crops

swimming holes!

hmmm. menopause?

yeah. ok. watermelon!

that last glass
of wine :-)

plastic bottles ...?

cows!

springs. wells

gardeners.
hobby farmers

the kitchen tap?
(well, maybe.
in some places)

air travel!
fourwheel drives!

Waste! Denial!

I WIN!
I WIN!

'Wet' is finished. The crowd claps politely, and as usual,
sets about figuring out how to cope.

JACQ CARTER

If I Am the River and the River is Me

Composed for 'The River Talks', a call to action to support the rejuvenation of the Ōmaru River in Glen Innes, Auckland.

If I am the river and the river is me
then for years I haven't flown so easily
in fact I sit and am sedimentary
and the paru has stolen all my clarity

If I am the river and the river is me
then few of us actually know my history
Ōmaru could mean absolutely anything
making me just a dirty stream

If I am the river and the river is me
most of my people have forgotten me
and they and those who live nearby
throw rubbish into my blinded eyes

If I am the river and the river is me
my purpose has been taken from me
it is not safe to drink from me
nor fish or wash or swim in me

If I am the river and the river is me
what future do I see for me?
Pollutants and chemicals still spilling into me
or somehow yet the cleansing of me?

If I am the river and the river is me
I hope that today you do more than remember me
and our words and dances and our songs
will see action right these many wrongs

If I am the river and the river is me
I pray my health will be restored to me
so I can again be a source of life
to all of those who seek my side ...

JONATHAN CWEORTH

Ghost Stoat

Winter lets you in
over barbs gentled by ice
past the wire's frozen pulse
you snowstep through the fence
unopposed

you elect yourself President for Life
in a secret ceremony
wearing a korowai of
pigeon feathers
fresh-plucked

you control it all
grip the dawn chorus by the throat
set the tūī gagging on loyalist anthems
dispatch a planeload of kiwi refugees
to seed other sanctuaries with fear

your enemies lay out poison buffets
soft beds lined with pheromone-laced straw
but you are past desire
a connoisseur only of power
and the purity of loneliness

in the cold vigilant nights,
you coin collective nouns:
a silence of mice
a nullity of rats
a void of possums

as you haunt your realm
a single lens records

JONATHAN CWEORTH

        your spotless coat
        shining eyes
        pawprints in the snow.

LYNN DAVIDSON

Speaking to the Otter

It doesn't break the water to emerge, rather
lifts water into otter-shape.

When one makes itself from river in front of me
I say *hello*

—the word formless in air. But oh the need to speak
because I am human, immersed in time, and this creature is fleet.

DAVID EGGLETON

A Report on the Ocean

you want to strip the atoll,
drag it all underwater,
you want to extend your tidal reach,
you want to bring the standing wave ashore,
darker tinge of your deeper waters
lapping from crystal shallows and aquamarine,
where roots of mangrove forests
bend like limbo dancers
beneath flow of warm currents

*

you survey, you eddy,
you search, you surround,
you lift the copra freighter
from its rusted anchors,
you drown the taro plantation
in its flooded salt marsh,
islands boggle and settle to your brackish surge,
the niu falls from the coconut tree
and floats out in search of another island

*

you leave your message
in anger at the bigger breach,
while buoys and fuel drums swirl
with bottles, toothbrushes, plastic bags,
cigarette lighters, tampon applicators,
plastic six-pack beer can holder
wrapped around muzzle of the dolphin,
driftnets in mazy patterns of screen-savers,

factory trawlers that vacuum shoals of fish
through washes of dead water

                *

your weather patterns of wild indigo,
your blue starfish, your purple thunderheads,
your forked stabs of lightning,
your hammering rain, teach
and tease in lagoons of your latitudes,
your guano islets lie abandoned,
your powder-white sandbanks glitter,
coral skeleton reefs fall away to the sea floor
from languid lisp of your breakers

                *

above you, bony ribs of thin clouds hang,
crossed by vapour trail's streak,
planet smudged to high heaven by carbon,
but, colossal from your horizon,
climbs sun, and the frigate bird glides
over the the shining mud, the living crab,
the octopus squeezing through rocks,
the parrotfish that revels in gentle rills
from big waves that undercut
the low-lying road and shrinking beach,
where your tide-beating heart rolls

RANGI FAITH

Losing our Mana

Time was when a foot
would slip on one
under the river bank
as you stepped into
the black water,

now they say there
hasn't been an eel
caught
for the last three tangi

and the brothers
down the river
come out during the night
& cut the nets—

they have to do that,
there is no kai
in the river,
we are losing our mana.

FIONA FARRELL

Eel

my youth was glass
pip of my heart
threaded
on gut and vein
for all to see

dark currents bore
me west then south
to a place where waves
shattered at a wall
of grey shingle

I wriggled through and
dropped into my life

bird pipe
flax rattle
mud suck
green leaf
spinning on water

suspended in my small
pond I lived my hundred
years forgetful of the sea
beyond the bar knowing
only the dimple of rain
soft blur of stars

growing thick as your
leg on shreds torn from
dead sheep snapping
at flies but never taking
proffered bait

I have lived as you have
lived: cautiously

but now I am old
and the sea knocks
at my head and there's
a taste to the water
that was not there before

I cannot eat cannot settle
guts shrunk to dry rattle
I turn head on to the current
and swim against the stream
drawn by the sound in my head

my eyes see more clearly
than they have ever seen
they are rimmed with blue
so that I may see in the dark
that lies ahead

I think more clearly
than I have ever thought
my brow flattens so that
I may move without impediment
through the dark that lies ahead

my belly is heavy
frilled with eggs
20 million strung
on velvet

I am become lean
and full of purpose

I cross the bar
on a moonless night
skin scraped blood raw
on sharp shingle

I drop back into the dark
into the ocean where
everything moves faster
and the lights confuse

I find my path my body
freighted with millions

I am heavy with the
future I bear it along
the dark path through
forests of kelp and
booming cavern
following the taste
in the water
and the stars marking
sharp left and right

I swim north then
east one undulating
muscle one blunt head
barking at the moon

I swim to the place
where it is time to burst

I heave and writhe
torn flesh

egg dances to sperm
the water glitters like
broken glass

and now that's done
I drift upon the surface

empty

old bag

skin for gulls

old bag

SUE FITCHETT

Story Lines

(A playful nod to erasure poet Mary Ruefle)

1.
pete takes us out to a tourist spot

pete's leg is falling apart
it's the glue

yet he drives us all
the way to Kaiteriteri

i point to those coastal mansions
mention rising sea

there's no climate change says pete
damn lies, a myth, hysteria

our car swerves round a corner
i sway in the back seat

next year pete's leg will be
reset with a stronger fix

if it starts tingling again in the future
the doctors will say *oh it's the glue*

pete believes in doctors
but scientists

that's another

2.
pete takes us out

pete's leg

      drives us
      to Kaiteriteri

      rising sea

      swerves round a corner

in the future

pete believes

3.
    takes us out

pete's   falling apart

hysteria

in the back seat

    tingling

another

4.
   take   out

          those coastal mansions

sway

5.
   akes

us

a myth

ALISON GLENNY

Candle

It was the year the city discovered gaslight. Soon, the footsteps of the lamplighter. Yet the darkness in the room when he composed the nocturnes was notoriously difficult to disperse. They had already burned the nightingales in the hope of immortalising their song. Now, as the ceiling grew luminous, he called for more night. Exhausted candles with blackened wicks slumped in pools of waxy light.

1. He gave her a feather belonging to a rare New Zealand wattlebird, and she wore it in her hair. But was this a sign of mourning, and if so, what had been lost?
2. The use of candles to illuminate the folios makes it necessary to consider the field of production, also a lingering attachment to nightingales.
3. If only the long, curved beaks of the females had not resembled ivory, sparking an immoderate desire for brooches, adorned with tiny chains. The last pair of laughing owls were sent as a gift to a collector, and vanished in the darkness of the museum.

JORDAN HAMEL

Te Aro

The city's history dissipates before me
villas retreat into the dirt
street signs run into puddles
I splash in their instruction as I walk

                       Rimu and Kauri sustain rapid growth spurts
                                  elbowing for position
                        pulling Carbon from the atmosphere
              harakeke flax chokes cracking footpath

sometimes I don't fear
the tick of eternity
after exhalation
maybe I just forget

                              there's an app that gives
                                you daily reminders
                                  you're going to die
                      critics are still out on its benefits

I want to unlearn the use
        of every natural resource
     strip the walls for precious metals
  bury them in the earth's fat
              stitch the wound shut

my bones   stole words for   'home'
   from   every   other   language
now dislocation runs   through me
  I wade   in   melted   pavement
fossilising that   which was   always
       here
               watch your step

DINAH HAWKEN

Losing Everything

I am the beneficiary of injustice.

And of Jane and Joseph's labour
and their love.

While they 'did something for a family'
there is no harbour
—we are tainted by confiscation.

After six generations, Tītokowaru
is still alive—as he said he would be—
in the general air, and in my mind.

      *

Some say that Jane and Joseph
'lost everything.'
But they began again:
they had their children
and they had land.

In the end
benefit built on benefit,
it became a solid dwelling
then a villa. It fed eleven children:

Ada, Ella, Nina, Mabel, Theo, Jessie,
Cecil, Oswald, Cresswell, Rose, Etta.

It gave them education and security;
a place in 'the better life' and the new land.

REBECCA HAWKES

The Land without Teeth

the year my body learns about want
I wake with hunger before the dawn
& rummage stale raisins from the muesli box

I suck them until the grapes rehydrate enough to denude
from their tasteless skins with my tongue   it makes them last
I pocket five withered fruits as though that breaks the fast

I whistle the dogs to me from their kennels
ready for ascension in corduroy & gumboot
& scuff through hoarfrost to the bush gate

up a route I take every morning in the hours before light
early riser clambering the crest of the mountain
away from the family home   the little cat warming my bed

the taunting kitchen   I come here
to get back to the stone that made me
to escape my mulchsoft body   returning

to the old volcano's swollen belly
fleshpink rhyolite muscular with its red web of veins
like mine   thinblood anaemic   red knees skinned on the rock

crumbling iron & crystalline structures
walking the dogs up the mountain in the dark
& stumbling my malnourishment but oh the sunrise

gilding the frost   fenceposts polished to silverware
the light so cold & loud it clangs like a cutlery drawer in anger
my pores whorl open into spiracles that gasp for extra air

becoming so unbodied   I sublimate
I get with the dirt
dizzy with cold & lightness I

am the black beech & the red tussock & filigree
lichen I'm honeydew I'm
braided into silver ribs

like the river below oh no I'm
more slender than a harebell stem
that can hardly hold up its pale head five petals

bluelipped   but this body's no alpine native
don't mistake me for some mystic nationalist
in my whiteness   & my rhetoric of wilderness

I am introduced to this place
like the gravel track tattooed on the slopes
& the unstrung barbed wire

spangled with needlefrost
I am   the pregnant heifer gouging out the valley
for fear of the dogs that chased her there

& I'm the dogs laughing for meat I'm buttery
gorseblooms yes I'm that
impregnable alien

pricklebitch skeleton
asway underneath my nodding
marzipan-scented yellows

forbidden stellated spindle I could rag you like plasticbag
all claw & fang   tell me I am your favourite coloniser
besides you couldn't unroot me if you tried

which I would know   I try so
hard my seeds lie dormant
still   I'm evergreen

me & my entire invasive species
consuming this landscape once so toothless
its homely fatbirds plumped defenceless in the shrubcover

where my little cat   collarless   devours them
with malice   leaves them gutted in my bed
as though I need a reminder of bones

HELEN HEATH

The Anthropocene

circa 2016

i. The phone that sounds like a ruru
I am walking up Aro Street, on a late spring evening
between pools of lamplight and darkness, well, as dark as it
gets in the city. A ruru calls, I check my phone for messages
—nothing. A ruru calls, I check my phone for messages—
nothing. A ruru calls, I check my phone for messages—a
ruru calls. I put my phone back in my pocket. The warm
night air presses against my face, insects in lamplight dance
their version of a mating call. Are all songs and dances about
reproduction? A ruru calls, the night air presses.

ii. The tūī that sounds like a phone
In 1983 my friend's parents got a new push-button phone.
It was beige, like my dad's Stubbies, with oval buttons.
Over time the plastic discoloured in the sun in their front
entrance. My friend spoke to her boyfriend for hours. But the
point that I'm getting to is the ringtone. Our dial-up phone
was heavy and its ring was made by actual metal bells hit by
a metal hammer to create a harsh bbbringg, bbbringg you
could hear throughout the house.

This new phone made a soft digital trill—brrip, brrip—a
higher pitch and softer, easily muffled. Until then I had only
seen push button phones on American television shows like
Diff'rent Strokes where the rich white daughter had her own.
I think she also had an electric toothbrush. It had been a real
sign of ostentatious wealth but then the push-button phone
came to regular New Zealand homes.

Anyway, the point I'm getting to is the ringtone—that soft but insistent brrip, brrip—I heard it today, 30 years later in the song of a tūī outside my window. A song that could never be answered by me, only another tūī, with the same ringtone, creating what feels like an infinite calling loop.

Did a tūī learn this call in 1983 and pass the song onto its descendants over the last 33 years? I imagine the slow progression … Or is this a more recent acquisition, learned from a nostalgic ringtone on someone's cell phone last summer as they walked home from the train station? I don't hear birds singing a telegram.

There is a David Attenborough clip on YouTube, from his *Life of Birds* series, of a captive lyrebird that mimics a car alarm, a camera shutter and a chainsaw to perfection. The lyrebird's syrinx muscle is the most complex of the song birds, giving them unmatched mimicry ability. Constantly adding to a wide repertoire has always ensured evolutionary success. Will the lyrebird still make car alarm sounds long after cars as we know them cease to be made? Or will it mimic new noises made by new technology that we haven't yet invented?

iii. Dawn chorus
All birds have their own place in the musical range of dawn chorus so they can hear each other. They maximise their singing effort by using different patterns of vocalisations, slightly different frequencies, and different timing. Birds that sing at nearly the same frequency, like tūī and bellbirds, will often alternate, with one bird waiting until the other is finished singing before he starts. Tūī songs range right across the frequency spectrum, so they need a bit of acoustic space and are usually the first birds singing in the dawn. When bird calls are lost from the chorus through extinction, other

birds expand their range to fill the gap. There is a theory that
mimicry may help a bird, and its offspring, avoid predators.
Will human sounds fill the gap in the range that we're
creating, with birds singing the chainsaw that cut the gap in
the forest?

iv. Who are you, huia?
Huia became extinct before field-recording technology
was invented, but a sound fossil remains. Tangata whenua
learned to mimic the sacred voices, to lure them into snares.
This fossil was passed down between generations, even after
the huia was gone. In 1954 Henare Hāmana was recorded,
whistling a huia call. Listen now, to a dead man calling a
dead bird, an echo from a machine.

v. New dawn
A crying baby. A cough. A cellphone. A camera shutter. A
chainsaw. A whistle. A car alarm sounding, sounding.

This poem references J.W. Bradbury and S.L. Vehrencamp's *Principles of Animal Communication* (Sinauer Associates, 2012); B. Krause's *The Great Animal Orchestra* (Little, Brown, and Co., 2012); and the article by Bryan C. Pijanowski et al., 'Soundscape ecology: The science of sound in the landscape', *BioScience 61*, 3, 2011, pp. 203–16.

JOHN HOWELL

Old Bones

High tides seep into the garden. Again.
Too high. Too often.
The fence crumbles.

Bleached coral stalks the reef.

The Tuvalu Rev pours, at his sanctuary,
his stormy outburst to his ancestors.

The cyclone has wound the clock to breaking.
The graves are waterlogged canoes.
On higher ground he re-inters.
The benediction blown to the oceans.

GAIL INGRAM

Recipe for a Unitary State

Take those brown hills, lumpy
with glacial form, strewn with
ancient herb and kettle lake. Add
merino for a living, some rabbit
& stoats for sport. Let stoat prey
and rabbit proliferate.
Introduce calici, a beneficial
virus—sure to choke off
the excessive taste of rabbit.
By now the herbs will have reduced
to hieracium and dust. Pour in
a cow or two along with most
of the braided river. It will
reinvigorate the capital
gain. Don't mind the extra nutrients
in the run-off—what you lose
in black stilt, you'll gain in the creaming.

You'll know it's done when it has reached
a smooth grassy consistency
with no hint
of all that vexing tussock.

KEVIN IRELAND

Moruroa: The name of the place

It is only a small island
but a name sticks to it
with as many syllables

as a continent. Names
have nothing to do with size,
they are tags to remind us

of the shape of our world.
It is dangerous to forget them.
We cannot retrace our lives

and our long voyagings
without words to light up
maps in the brain,

though sometimes names
may also echo the sound of a place,
catch the graunch of a glacier,

the wind that sucks up
the sands of a desert,
or the slow rumble of tropical seas.

Names radiate energy.
They give tongue to triumphs,
pleasures, desolations.

They may taste of syrup
or salt. They can lie to us
or cast stones. The syllables

may even suggest
split fragments of meanings.
They can trigger reactions

that detonate beneath oceans
of memory, then crumble away,
falling in on themselves, leaving

only their corruptions eating
into the reefs below the waves.
Names can break bones.

ANNA JACKSON

Huia

Huia feathers were always rare treasures,
kept in waka huia,
treasure boxes.

An iridescent bird, blue-black like petrol,
with a greenish sheen,
rarely seen,

the huia hopped along the ground, grounded.
But sang like the tui,
like a flute.

Dressed in treasure too valued by people
for the bird to be valued as a bird,
the huia is no longer heard.

When the Duke of York was presented
with a huia feather for his hat,
trade

in huia feathers leapt to extinction.
Now the waka huia preserve
other treasures.

This is my waka huia for the bird.

ASH DAVIDA JANE

2050

Carbon capture may indeed prove to be 'magical thinking', but the cruder technologies—we know these will work. Rather than sucking carbon out of the atmosphere, we could shoot pollution into the sky on purpose; perhaps the most plausible version involves sulfur dioxide.—David Wallace-Wells

I pay daily attentions to colour
                7am waiting at the bus stop under
                        a sulphur-red sky
       burnt at the edges where it
              sticks to the horizon
fading to a midday dull white sheen

        the ocean a room of
mirrors reflecting itself
        the edges of waves tinged pink
              like we're on another planet
but we're exactly
        where we've always been

except there's a PE teacher
pushing us to go faster than we want to
                jogging into an apocalyptic future
                      in polyester shorts

what if kids dare each other to stand
                      outside in the rain
knowing the trees burn faster now
        having watched a video at school
                of red flames on a red sky
        shot in slow motion

what if   above our heads
we project squares of blue
with white blurs passing across

something missing at the core but
        the absence imperceptible
to the human eye

I could count the shapes as they pass
        I could
come to love them
        against a false sky
I'm afraid
        I could forget
there was ever another way

TIM JONES

All That Summer

All that summer we sailed the drowned isthmus,
Miramar Island bulking east. Diving
was an anxious wait for murk-filled water

to yield its occasional treasures, relics of better days
left behind as the frantic dikes were overwhelmed.
Out by the drowned airport runway,

the never finished extension lost beneath us, we faced
long rollers carrying Antarctic meltwater northward,
braved the sudden southern chop and squall

to plumb abandoned warehouses, corroding cars.
So many days we returned empty-handed
to the boatshed on the Wadestown shore,

worked the elaborate locks with reddened fingers,
climbed the hill to short commons and mixed
parental signals of frustration and concern.

It was a life lived in increments of bad news, a
Government of bluster and paralysis, its authority
manifested in chain-link fences and pronouncements

no longer listened to on matters that concerned
only those sited most securely inland. At the water's edge
the social contract washed away, replaced

by alliances more fickle than the weather.
And the sea still rose, icecaps converted to ocean
by generations of accumulated arrogance.

That was all before our time. What we knew
was the rising wind, swoop of storm,
slack and snap of sails, one of us waiting aboard,

the other diving the ruins of lives lived
in those final glittering years of denial
before the ocean washed all doubts away.

ANNE KENNEDY

Flood Monologue

You never discussed the stream
and no doubt the stream didn't want

your discourse (its own merry way)
but now that you live by the stream

a mosquito has come up the bank
and bitten you, and the stream

is in your bloodstream. You buff
the site of entry like a trophy.

Your chuckling new acquaintance
takes your cells out to the sea.

*

It goes all night, you tell your friends
drinking wine to warm the house

(already warm), and laugh of course
like a drain. Later in your roomy

queen you listen to its monologue—
ascending plane that never reaches

altitude. Your fingers stretch
from coast to coast to try it out,

this solitude, while water thumps
through the riverbed.

*

You're not exactly on your own.
Teenagers come and go, the screen-door

clacks, cardinals mob a little temple
hanging in a tree. A neighbour with a bag

of seeds asks you if you mind
the birds. There is that film, and the flu,

but no. In the mornings earlyish
you slide the rippling trees across

(Burnham Wood) and watch
six parrots lift like anti-gravity.

*

At sunset a rant about the dishes—
you've worked all day, unlike

some people! The tap runs. The sun,
tumbling over Waikiki, shoots through

the trees, gilds the stream (unnecessary),
stuns you in the empty room. Every day

for ten years (you realize, standing there)
you've crossed the bridge etched Mānoa

Stream, 1972, back and forth,
except the day the river rose.

*

Some facts: mongooses (sic) (introduced)
pee into the current, plus rats and mice,

the stream is sick. All the streams.
Mosquitoes—your messengers and those

that bit the teenagers whose young blood
is festive like the Honolulu marathon—

could carry West Nile virus. Often fatal.
Probably don't, are probably winging it

like you, and you will go your whole life
and only die at the end of it.

*

The stream doesn't *look* sick. It takes
a pretty kink near your apartment.

The trees are lush and spreading
like a shade house you once walked in

in a gallery (mixed media). The water
masks its illness like a European noble

with the plague—a patina, and ringlets.
You're pissed about the health issues

of the stream, and healthcare, because
it has your blood, you have its H_2O.

*

You think it's peaceful by the stream?
Ducks rage, waking you at 2am,

or thereabouts. Mongooses hunt
the duck eggs, says your son. Ah, you say.

That night the quacks are noisy, but
you fret in peace. Sometimes homeless

people sleep down by the river bank.
Harmless. One time one guy had a knife.

They still talk about it and you see him
ghostly like an app against the trees.

*

All your things are near the stream,
beds, plates, lamps—you're camping

apart from walls and taps and electricity.
Your laptop angles like a spade,

and clods of English warm the room
(already warm). They warm your heart.

Overall you have much less, because
of course—divided up. But you're lucky

or would be if the stream was squeaky
clean, and talked to you.

*

The stream had caused a little trouble
in the past, i.e., the flood. Not its fault.

900,000 people pave a lot, they plumb
a lot. Then rain like weights. From a safe

distance (your old apt) you watched
your little water course inflate and thunder

down the valley taking cars, chairs, trees.
You saw a mother and her baby rescued

from a van—a swimming coach, with ropes—
the van then bumbled out to sea.

*

One apartment in your complex
took in water in the flood. And mud. It was

this apartment. You've known it all along,
of course, because you watched.

They fixed it up. Lifted carpets, blasted
fans for a week. Repainted.

It's pretty good. The odd door
needs a shoulder still. In certain lights

though, on the wall, a watermark,
the stream's dappled monogram.

*

You're talking clichés—water under
the bridge, love letter from a lawyer,

serious harm, sunk without you.
The stream has been into your bedroom,

and you in its. Remember reeds, coolness,
summer afternoons. You loved

the stream. Its stinging waters send
a last message in lemon juice:

fevered me, infected,
flooded me.

If I'm fucked,
you're coming with me.

Sincerely,
the stream.

ERIK KENNEDY

Phosphate from Western Sahara

[*New Zealand fertiliser companies are the only significant remaining foreign buyers of phosphate from Moroccan-controlled Western Sahara.*]

It travels along the world's longest conveyor belt
surrounded by the world's longest minefield.
Each mine says, '*My* field, *my* field, *my* field.'
It travels to the land that fertiliser built.
Sorry, Western Sahara, New Zealand needs this.
Phosphate is purveyed like a louche commodity, like a white kiss,
like a lost memory of self-determination.
Two peoples trapped in a laughable half-rhyme: Sahrawis, Kiwis.
It remains our position that we are operating within UN expectations—
this is the accounting software justification
of abstracted minds, which apologise and at the same time race
to take what a dying industry makes. Thanks, free marketeers.
'Peak phosphorus' in thirty years,
and until then this comedy carnival ride of equivocations and rocks.
You can see the phosphate dust from space,
like a tantrum in a sandbox.
You don't have to look hard for motives
when someone guards their shame with barrenness and explosives.

MEGAN KITCHING

I. Pūhā

    (Sonchus kirkii)

Growing with abandon in blue-grey swathes of before
on these coasts, heads in the clouds, he ao, he aotea

while on seas dark-dyed with trade men mark up
where the world begins and each thing circumscribed within.

    It grows and we pick, we rub and we cook, we eat
    and laugh and shit and it grows and we pick. We don't need

    your books and your lists. Wouldn't you rather hear
    what common-or-garden gossip's brewing?

All the same it's drawn out of the ground into the salts and soot of ink
the supple remembrance of the page, the pose, that bristly jizz

which the tangata whenua know: tender, stringy greens with mussels
hot stone steamed even before the ships sail in with their pigs.

    Sow thistle gets under our boots, we diggers
    of slip and gully seeding fast as we can clear. How like home

    the native can look. But a weed's a weed and we can leave
    chaps like Kirk to sort one from the other.

Now the endemic strand clings to capes and cliffs
    where scientists stoop to record and it's the Pākehā
pūhā we spray or eat for old times' sake, a bitter reminder.

LEICESTER KYLE

from **Death of a Landscape,**
An Aside:
Advice to the Rehabilitator

(conditions permitting—money, staff, skills, the weather)

Be inclined

Put the granite on the bottom
sandstones above
on the top Kaiata

and get the drainage right
with nothing left dry—
north to the Patrick
south to the Waimang
with no acid

Then the plants in their groupings—

                                                  eleven

Don't miss a thing—
it might be a key
and they all fall down

There's the little plantain
and C. parva (?)
Put them in their places
and keep to their kinds—
   1.   Rata, kamahi, and mixed beech forest
        Nothofagus solandri var. cliffortioides
        Halocarpus biformis
   2.   Mountain beech and podocarp forest
        Eleocarpus hookerianus & Gahnia procera

3. Mountain beech and podocarp scrub
   Halocarpus bidwillii and Pseudopanax linearis
4. Manuka, wire rush, tangle fern and red tussock
   Empodisma minus, Gleichenia dicarpa
5. Manuka shrubland and scrub
   Leptospermum scoparium & Phormium cookianum
6. Manuka tussock shrubland
   Chionochloa rubra & Dracophyllum palustre
7. Sandstone pavement vegetation
   scattered shrub, rush, tussock, and herbfield
8. Disturbed or bare ground
   (vegetation almost absent)
9. Red Tussock Grassland
   Lepidothamnus laxifolius & Myrsine divaricata
10. Red Tussock and Mountain Flax grassland
    Epacris pauciflora, Lycopodium fastigiatum
11. Tussock herbfield
    Oreobolus pectinatus & Carpha alpina

You know the plan—
    Once the plants are back
    the birds'll come,
    the gecko too and the snail

    then those things that haven't been discovered

Give them a place in the sun
if you will
so we might know them

ARIHIA LATHAM

Birdspeak

Ruru, koukou, your call
a kōauau

Your name,
the only taonga of you I have

Place it gently on my tongue
let me speak as if a bird

Let me speak of you in our reo as if
your memories have wings

Populate my mouth with the notes
of your feathered song

Koukou, my great tāua
were you named for your singing

Or your lament
we have sadness

Rife in our lines
did you call to the depth of the night

Did your sharp eyes hunt
the light from the moon

I hear the ruru koukou at my window
it has never visited me in this city before

My baby kicks at my lunar belly,
both wide awake in the dark

Koukou, you call to us
name the pēpi for every pause in our connecting breath

Let it be an oriori
for her memories

Let it be the sound we speak
as if our mouths were curved

Beaks to hold the vibrato
through all the strange noises

Our life is filled with
your call to us, a kōauau

HELEN LEHNDORF

Oh Dirty River

The town where I grew up
was small, ugly and smelled
like burning blood.

Most of the dads and
a lot of the mums and
heaps of the big brothers and sisters
worked at the Freezing Works.

Thousands of cows and sheep
and even a few hundred pigs
would get trucked in, slaughtered,
chopped up and packaged
in cling film each day.

The burning-blood smell
came from the incinerator
where they would burn
the bits left over.
Though, some of it got pumped
right into the river
which ran through the town.

In our town,
people called the works
'The University'
because it was where most of us
ended up going after we left school.

People also used to
call our town 'Lavender City'
because of the burned-flesh stink.
So you can't say
we didn't have a sense of humour.

Yeah. You could make a joke
about it. But only if you're
from there, eh?

Otherwise, you're just
getting smart.

TOI TE RITO MAIHI

Korari / Harakeke

The scent and sound and touch and sight of you
is part of my being.
Without you I am incomplete,
vulnerable ...

But
when we are together
your diversity, your complexity,
excite and entice
towards a merging of self and plant—
a melding of the mauri of both
in a voyage of respect, of intrigue,
of speculation, of creativity,
wherein another possibility forever beckons—
my fingers following multiplying images
within my mind ...

BILL MANHIRE

Huia

I was the first of birds to sing
I sang to signal rain
the one I loved was singing
and singing once again

My wings were made of sunlight
my tail was made of frost
my song was now a warning
and now a song of love

I sang upon a postage stamp
I sang upon your coins
but money courted beauty
you could not see the joins

Where are you when you vanish?
Where are you when you're found?
I'm made of greed and anguish
a feather on the ground

*

I lived among you once
and now I can't be found
I'm made of things that vanish
a feather on the ground

SELINA TUSITALA MARSH

Girl from Tuvalu

girl sits on porch
back of house
feet kicking
salt water skimming
like her nation
running fast
nowhere to go
held up by
Kyoto Protocol
An Inconvenient Truth

this week her name is Siligia
next week her name will be
Girl from Tuvalu: Environmental Refugee

her face is 10,000
her land is 10 square miles
she is a dot
below someone's accidental finger
pointing westwards

the bare-chested boys
bravado in sea spray
running on tar-seal
they are cars
they are bikes
they are fish out of water
moana waves a hand
swallows
a yellow median strip

moana laps at pole houses
in spring tide
gulping lost piglets
and flapping washing
girl sits on porch
kicking

CAROLYN McCURDIE

Ends

1.
I remember how he used to be funny,
that cartoon man on the street corner, wild-eyed
and lonely with his placard: *the end is nigh*.

Now here's the data. Here are the graphs.
Here are the aerial shots, the cracks
in the earth, in the ice.
Here are the first refugees.

Storm coming,
storm beyond our knowledge of storm.
Grief coming,
beyond our knowledge of grief.

2.
Slam the door. Pull the blinds.
Like a child, I dive to hide under the bed,
and find you there, already hiding.
You move, to make room.

You do this revolutionary thing—
you move to make room.

From your pocket you take out a square
of squashed chocolate.
Trade Aid, you say. You break it,
give one half to me.

You do this revolutionary thing—
give one half to me.

Warm melt in our mouths
as we talk about storm, grief,
our own jelly-livered weaknesses,
and yours, it turns out, are as bad as mine.

We do this revolutionary thing—
we tell the truth, we listen.

And you pull out your phone.

3.
Your phone—
a small crucible
that spills the world, its joy,
its stink, into your palm.

And I sink back to despair.
Because, there on your screen, the trolls,
the sad addicts of power and wealth,
who think they inhabit
some world that's not our world. Not Earth.

But, look. You scroll over continents,
islands, and the people, thousands of people
on beaches, in parks, in stone streets
hard with history. They're chanting.
They're singing, drumming, in oceans of banners,
placards held high: *the old story ends*.

What d'ya know, you say:
we are millions.

4.
We crawl out from under the bed,
crumpled, sticky. We have cobwebs
in our hair and we laugh
at how inspiring we're not.

We stand at the door, watch the wind
lift thistledown, sunlit, buoyant,
into the sky. Birds twitter on the roof.
Then we know—

song coming.
Song coming, beyond any
we've ever before lifted in song.

MARIA McMILLAN

How They Came to Privatise the Night

It began with shadows
Our dark selves
Small nights we carry with us
Stretched and shrunk
Rushed into corners.

Striding into the sky
Like the Chinese lovers
Whose bridge is the Milky Way—
Distance was nothing to them
Or waiting seven years.

Clearly of private benefit
They said: The shade they offer.
The company. The sense of self.
Hitherto pricing has not reflected
Their true value.

*

Dusk was much the same.
A wilful resistance
To applying the forces
Of the market.
The stillness.

The nuances of colour.
The way mountains seem closer
And the white houses
On the hills of the city
Shine like angels.

*

Then night.
By the time we heard
The sun had slipped between
The South Island and the sea.
Gone like music at a party
You are walking away from.
Night was a business.

The government maintains
A regulatory role.
At the end of every street
Yellow jacketed officers collect tariffs.
They watch for you.

Watering the garden
In the coolness.
Talking in quiet voices
On the porch
Inside the kids dream.

Letting the cat in and out.
Opening the curtain to sneak
A glimpse of the orange
Mouth of moon.

Functions are contracted out—
Absence of light.
Comfort to the weary.
Frost. Fear. Astronomy.
Navigation. Romance.

The dark profusions of freesias
Letting go of themselves.

CILLA McQUEEN

Frogs

The atmosphere is thinning—
the world is getting dirty

as the outer epidermis eats itself.
The frogs are vanishing.

Who will recall in Costa Rica
the webbed feet of the flying tree frog

filling like a parachute as it soars in the trees?
In Chile the four-eyed frog with eyes on its rump?

Gone without trace that drab frog
that flashes repellent patterns,

the Madagascan frog that turns bright red
and puffs up like a tomato?

Gone the ivory frog of the arum lily
that turns brown to match the dying blossoms?

Silent forever!
Who will remember

the bong-bong banjo
call of the pobblebonk?

VAN MEI

There's Real Mānuka Honey in Heaven

In the year 3000,
lizards will be the last survivors of the remaining world order
but it doesn't matter 'cause we're all going to die!

in twenty-two years i'll be hitting my mid-life crises
as the Amazon finally collapses from heartburn
a wizened crone releasing the IV drip
from glazed root veins

a global conference of bees will be livestreamed strapping on
army helmets khaki stripes and matching jet packs
then flying off into the stratosphere in tiny astronautical booties

the bonnets left behind had their URGENT: agenda emails
marked as spam but who cares about spam
when we'll live underwater and eat raw salmon every day
and if we're poisoned then we'll relocate

leaving the planet, it's one small step for man
as all the mammals left on Earth's Ark of the Covenant
study *Home & Garden* magazines as textual predecessors
that foreshadowed our ecological demise
under the watchful eyes of agricultural industries

while the tuatara sings a eulogy to the end of the anthropocene
five hundred fortune cookies crumbled to ash
as the cockroaches wait for spring

KARLO MILA

Matariki: A call to kāinga

Kāinga, 'aiga, 'āina.
Let the morning mist
remind us
with its cold, fresh breath,
that we are alive.

Matariki:
marked by a constellation high in the sky,
barely visible to the human eye,
known to collectives of seekers
searching for signs,
lashing ancient
markers
of time.

Can we return to rhythm?
Remember lives lived along the arc of a
wiser calendar?
When we were allowed to wax and wane
with life itself?

Matariki, can we stop?
Our fossil-fuelled forever forward,
full speed towards an irreversible end?

Can we take stock?
One million species endangered.
The ruin of rivers, fallen forests
the mined, the fracked,
a carbon-choked sky
holding its breath
for us.

We are a water world.
Oh ocean, we ask too much of you.
Gagged, plastic-bagged.
You hold too much heat
on our behalf.
Vast dead-zones
oxygen-deprived,
acidic tides.

Missing and murdered
beneath your waves.
Coral reef graves
point their bleached,
broken fingers at us.

No, tomorrow
is where we find ourselves
today.

Heads or tails?

If ever the world needed to come
full circle
it is now.

Matariki,
time to honour different ways
of seeing the same night sky.
Different ways of being human
and being alive.

Let lost languages
give us ears so we may hear,
eyes so we may see.
Hearts so we can translate

the thudding rhythm
of the co-created:
that which feeds us, breathes us,
the substance under our feet,
fresh water that flows in our veins,
air in our lungs,
stardust in our bones.

Let us call on our family of relatives
to help us heal.
Bright light of the stars,
fierce power of the tides,
enduring knowing of the stones,
cleansing of water that can still heal
everything it touches.

Then let us attune
beyond sound
to that most profound
to mauri/mauli—
the essence.
So we can stand in sacred places
that have plans for us
encoded into the soil of its soul.

Build the marae/malae
to house a future we can live in.

Let us not build meeting houses,
let us be meeting houses,
let our bodies
house the meeting
that needs to happen.
For we are out of time.

Mataliki, Matariki, Makaliʻi, Mataʻiki
Time to stop. To take stock.
Pause and reflect on all that we've lost.
To remember our dead. To shed.

End the aching arc of bitter farewell.
Find fresh.
Enter a new cycle,
following hard-earned,
hard-learned bends
of where we have been before.

Awaken.
Act.
Hope for another
turning turn.

HARVEY MOLLOY

Dear ET

The white dusting on our poles is melting.
Only bears and penguins live there.
Ice upon ice like a migraine in a stark neon-lit supermarket.
Ice like a sharp slap or a wake-up punch to the head.
Streams of melting snow fall down
shafts in the ice sheets,
which have their own name—*moulin*.

If you came when the plants grew the sky
or when the fish bloomed the sea
or when the lizards pounded the forests,
come now to see the short-haired apes go apeshit.

ELIZABETH MORTON

Birdlife in a Broken Century

I

I'm a delinquent, skipping the stony tenements
that belch smoke and invective. I leap TV aerials,
scuttle slow as thunder along the window ledges
ducking chickhood 'mares and cigarette ash tapped
from the bedrooms above. I'm a buzzard,
inconspicuous and dull-eyed. I'm a lifetime of birds.
I've been tarred and feathered like a snitch
and left for dead in the village stocks. Apples, stones—
the shrugged sadism of people who swat spiders,
whose barn animals know them by a blur of timber.
I could run into the fauna of gentler suburbs.
I could talk to the bobby calves waiting in the trailers,
moo soft milk vowels and hope the blood of one
does not rattle the breath of its brother. I am guilty
of queuing, shamefaced at the altar, begging for meat.
Forgive me; forgive my hands their collateral.
If I am the birds, I know just what I take.
If I am the sky, I hold the buzzards like my heart.
If I am a person, there can only be attrition and slick sawdust
between this thing and the next.

II

Everything I touch is flames. I knit heat to its animal.
I'm the drought of the leap year, the marrow lapping
in the dry bone. I could eat my own rind, lick the salt
from the edges of me, but I would not be sated.
At the altar, I fetch exoneration like a stick.
Here is the skillet and here is the heartbeat;
here is the iron and the winter and the hangover

that comes upon waking in a supermarket aisle
where no beasts bray. In the next life I am plastic—
tonnes and tonnes of clingfilm and polystyrene beans.
*Look, the mattress is on fire! Look, I'm on my tiptoes,
yet the ocean is over my head!* If I had a dollar
for every time I strike a match, I'd be a rich man.
As it happens, everyone I know is a pyromaniac.

III

Ma said leave only footprints. My carbon footprint
is a kilometre across and six miles deep.
Last week I made a nest with bubble wrap and hair
for all the kids I'll never have.
At the altar they say I'm a Zero but they do not see
the scope of my despair.
If I am a bird I am stuck circling the carnage.
If I am a person I'm sat in a burning car listening to radio songs
and dialling numbers of friends I don't see in real life;
not even the jaws of life will find me.

EMMA NEALE

Huia

Have you heard of the huia?
Yes, I have heard of the huia:
passerine, black with a green sheen,
or bluish, yes perhaps bluish,
and the black itself metallic,
or perhaps lake at midnight-ish,
perhaps like a tūī, a
large tūī with orange wattles,
but precisely what wash,
what quality of orange?

All comparisons we draw
fall now more approximate
than any historical simile;
the female's beak long and curved
like an upholstery needle
(we still have upholstery needles)
the male's short like a crow's
(how long will we still have crows?)
the tail feathers tipped with white
like the moustache of a hunter-collector
as he plunges in to the creamy head
of a victory pint of lager, but the greenish,
the greenish blackish feathers,
perhaps the exact blackish greenish
of a leaf of punga as it tilts
beneath the beech that would have been
on the deforested hillsides and gullies
in a certain slant of mist and sun
as time slides between seasons
but *piccolo-piccolo, piano-piano*,
strain for the melody,

what was that song, how did it go,
Uia, uia, uia,
where are you? Where are you? Where are you?

Only lyric mimicry survives:
a recording of Henare Haumana,
a 1909 Huia Search Team member,
as he whistles an elderly man's
childhood memories—
{ -.'-.'-.'? -.'-.'-.'?-.'-.'-.'}

Ah, how I want to hear Henare and the huia,
a warm weight in the throat
as if it were tuned
in perfect pitch
to expectation's A

{ -.'-.'-.'? -.'-.'-.'?-.'-.'-.'}

click, bash, click, bash, click, click
can the mind's ear come near to
{ -.'-.'-.'? -.'-.'-.'?-.'-.'-.'} ?

404
the page you requested is not available
the server encountered an internal error ()
that prevented it from fulfilling this request
you do not have permission to access these files
page not found

huia, huia, huia
uia, uia, uia

{   }

NAOMI O'CONNOR

So You Don't Belong, Pohoot

What can you do about it, eh tree.
Suck the water and see, maybe—
send your roots out through rock,
tie yourself in knots to the bones,
spread a blanket for summer picnics.
Act confident, act like you're blossoming
like you have a lot to offer. Look as good
as you can—grey green in the lonely cold,
scarlet and proud in the heat. Open up,
take each day for what it's got. Take
a leaf, unfurl it, take the next, unfurl it.
What do pieces of paper have to do
with you, a tree, alone on the shore
already burning?

KIRI PIAHANA-WONG

Piha

There is a small blue pot, filled with daisies
picked from the roadside, sitting on the
windowsill, framed by plywood, glass,
the dim, warm, pre-cyclone light
—it is mid-afternoon.

There are grapes, not yet ripened, hanging
on a trellis above me, the trellis covered in
clear plastic, giving the illusion of open
space, protecting me from the rain.
Behind me, pōhutukawa are flowering,
our brilliant red Christmas trees.

Because yes it is Christmas,
it is Christmas Eve, and this is where
I start to lose it, I stop looking and
start listening, I'm listening to you
drumming, Ahurewa singing, and
while I want to describe the precise
nature of the sound, what I can hear,
all I am thinking is

—nobody plays the drums like you do

        and then I'm lost, you see,
I want to be lost and I am lost
        and I'm   gone.

Sometime later I come back to myself
to the sound of flowers. I am in a high
place, close to the sky.

There are light green leaves above me,
as perfect as stencils. There is a creeper
growing, wrapping itself tenaciously
around the trunk, the limbs, of a pūriri
tree. I think the vine is winning, it is
smothering the tree, but then I see no,
the tree is still strong, although part of it
looks dead, and then I wonder if nature
even thinks like that.

And I'm a part of nature too, never more so
than now, this day, this pre-cyclonic post-
apocalypse-that-never-came rainy early-
summer Christmas Eve, this early evening/afternoon.

  There is more rain falling now.

      It runs in rivulets from the top of your head
  down the bridge of your nose
                  onto my half-open mouth, running over
my lips, and it runs over my chin and
                            it runs down my body and pools
            in my centre, and then as I turn over to
        press my face, my warm, bare face, against the
                        grass, dying leaves, the earth, I feel it
            —the sky's water: all the wondrous light
              weeping joyous tears of the sky god,
            Ranginui, running down my side and
      into the earth, Papatūānuku, and then

                                                settling there.

ROBYN MAREE PICKENS

Praise the Warming World (Try to)

—*after Adam Zagajewski*

Try to praise the warming world.
Remember the crisp delineation of seasons; the sting of winter
and the pond that froze each year, blades clinking on ice.
The pinecones that fell around the edges and lay entombed
all winter long and your mother who swept away fresh snow each day.
You must praise the warming world.
You watched icebergs shear off Antarctic glaciers
one of them floated off the coast of your city
while others melted into the scent of salt.
You've seen the refugees going nowhere
you've heard the silence of detention centres and deportation.
You should praise the warming world.
Remember when I licked warm honey from your burns
in a quiet room bright to the eyelid with sun.
Return to that small hut perched on the edge of the lake.
You traced ripples made by oars with your hand
and snow filled in the earth's furrows and sores.
Praise the warming world
and bare branches strung with bird song
and the stray light brushed with wing beats
on a clipped winter's day by the pond.

NINA MINGYA POWLES

Whale Fall

(i)

Last night you heard men shouting, swearing,
saying the whale had got free, the rope had
snapped and they had watched it swim off blindly,
suddenly bewildered at its freedom when
just seconds ago it was set to become the ribs
and handle of a lady's summer parasol.
They talked of going after the wretch today.
You remember your dream last night:
a huge whale swimming on and on,
holding its breath and going down deep,
deeper, far below the point where light
touches anything, where there are no words
fit for this kind of darkness, and this is where
it shuts its eyes and puts its belly to the seafloor
so that its skull might one day shelter sea
monsters or its bones might be mistaken
for some great shipwreck
but not for a thousand years.

(ii)

There is a phenomenon called 'whale fall' (n., singl.),
which is a dead whale that has drifted down to the abyssal zone,
the deeper section of the midnight zone that is never touched by sunlight.
The whale's skeleton, stripped bare of soft tissue, transforms here
into an ecosystem of organisms, supplying nutrients to a community
of deepwater scavengers over a period of decades. Whale falls are

normally located by sonar technology, and have been mistaken
for broken bits of airplane fuselage
and wartime submarine wrecks.

The place below the point
where whales fall is the hadal zone.
There is nothing here except certain kinds
of jellyfish and tube worms and the darkness
is only sparsely interrupted
by bursts of bioluminescence.
You can still see them when you shut your eyes.

VAUGHAN RAPATAHANA

Topside, Nauru

like one-armed
drownings,
these
scarred
escutcheons

crash
through the quaky crust,
dusty,

in no clear
pattern

some
     stalk
          others

a few
                    adrift,

even more
                                        abandoned;

all   eyeless.
doomed.

there is <u>nothing</u> else here

frigate birds stay away,

even the leaves have left

```
        gg
just ja^^ed
monoliths:
dead men.

expedition
            over—
stone surrender
to the indefectible
sun
```

RICHARD REEVE

The Old Breed

Schist-head, bog republican, right river ranker
at work on the slippery floor of their stream. So they regard him
who knock heads with this old outcrop

professor, ecologist-irritant to the corporates,
friend of a just farmer, foe of the plain greedy.

The Busters, clan Tuft-Giddocht, Smiths, Feckhoffs,
sniping at mountains, skinning the tussock
for their dairy and gondolas, vehicle-testing stations,
wind farms, reservoirs, quarries, skifields,

insatiable; yet he tolerates their ignorance.
Exhausted humanist, botanist at home on a scarp
assessing grasssweat, or yanking out the pines
that creep from plantations bordering the Takitimu Mountains,
the Southern Garvies, the Lammermoor;

shadowy hill beauty in his threats, the vulnerable

reluctance of water and earth to conform.
Unimproved country, who will fight for it
when time takes him?; when man worn down
no longer considers the gentian and the gecko,

and jackhammer, plank and crawler crane
fake grumbling congress in the hinterland.

TE KAHU ROLLESTON

The Rena

It began. As ....
the essence of death, itself sept, from the monsters depths, into the sand
impacting, all creatures, from the air, sea, and the land,
she was stuck!!
Between a *'reef and a hard place'*,
jammed
like boiled fruit pulp in a jar case,
her knife-like features with a sharp, blade-like base, pierced my Moana,
while oozing, and bleeding this dark paste,
she was stuck!!
No *anchor*, nothing *butter*, dark taste,
the volatility spread, churning my once bright pantry and sanctuary into
    a dark place.

We were going wild, lives spinning,
out of control, with the wild life, killing,
that was occurring, as the government sat around, downing their *Caésar
    salad* just chilling,
how dare you,
poison the swells and the realm of Tangaroa,
then sit around and *watch* ... as *time ticks on*, while doing nothing at all,
to those with the access and knowledge that's a food basket and store,
payments made, with the practice of Kaitiakitanga, Tikanga, and L.O.R.E
    Law,
until that day, when this *blanket* of *death lay*, on our sea *bed*, and he was
    almost a beach, dead, *for sure*,
I saw them, an army of Taniwha, surfing the waves,
in the shape of shipping containers, though nothing within them could
    be contained,
armoured in steal, stealing, taking the life of my Moana away,
unless you were raised, to be at one with the sea,
you could never see,

believe. Understand or feel this sort of pain,
while it was happening there was a culture clash,
Money vs Mana
who determines and measures success and wealth
for is wealth
the ability to be able to collect enough food to sustain yourself?
Or is the wealth
forever to be measured as the assets in your possession and cash in your
    bank account?
The sea ...... Well that's a *resource* to followers of capitalism,
but what's a *re-source*, to our *Mauri-source*,
my people's very essence of living.

As soon as it happened,
we were there, an army in gumboots and latex gloves that protected
    nothing believe me,
that's the sort of power and love, shared and felt between Whānau Hapū
    and Iwi.
To overcome my anger I had to find a silver lining,
and what I ended up finding, what a unity
the one-ness that can come from such a tragedy, and that's the only
    reason I'm still smiling ......

SAM SAMPSON

Erasure

history in a nearby tree
isolated sightings ~~(aligned by absence):~~   ~~huia extinct~~

for Don Binney, artist, conservationist

TIM SAUNDERS

Dad's Piece of Sky

My father's piece of sky
was directly above the maimai.
A thin strip that captured
clouds and weather,
framed by macrocarpa branches
cut for camouflage.

He would tip precariously backwards
on the old wooden ENZA box
and cast his gaze upwards,
so that tufts of white chest hair
escaped the throat of his Swanndri,
perched like a magnificent Tūī.

Raupō tapped gently
against corrugated iron,
a huntaway barked
somewhere over paddocks
and swallows threw mud
in the corner and called it home.

Dad would stare patiently
at that patch and wait for ducks
to drag Vs across the dam,
and he would tell me everything
I have ever needed to know
in his silence.

These days I know he can always be found
in that little patch of sky
fringed by macrocarpa
amongst the weather, the incessant
Raupō tapping, the dog's distant bark
and the swallow's fist-sized nest.

ELIZABETH SMITHER

Port Hills, Canterbury

Never such a violent declension
or shortage of words for it.
Not 'magnificent', far more slippery
as they tread with giant steps through gorse
or peer over a road they've made perilous

but on their edges softer trees, ridiculously small
try to pacify these giants with bibs
and smaller still, the close-placed cherry trees
signal like fairy lights and declaim their liturgy.
Hope, hope, hope along the edges.

RUBY SOLLY

River Songs – Waimāpihi

Pāpā, I've lived in this valley
ever since as falcons' prey
you dropped me in.

I hit the ground digging.
Sun piercing through
the cracks in the roadways.

This road runs deep here, Pāpā
and beneath it
the river deeper still.

I hear it rage under our atmosphere,
when I dream you
diving through concrete

with me holding your ankles.
They call this the Waimāpihi
after our tūpuna who bathed in these waters.

Ownership is different here,
it rises up from the ashes
of guardianship.

One night
I fall into her stream,
my body seizing

in an electric storm.
I feel her push up the concrete
that bars her from the sun.

Return to our valley, Pāpā.
You can have the bed,
I will weave a mat and sleep on the floor.

I will boil you potatoes and roots,
I will drink the broth.

I will pull the land up to meet you,
by pushing myself further and further into the sea.

You will walk a land as long as the eye sees,
while I rip down fences in your distance.

I will wrap you in blankets,
while I sleep pale and bare.

I will smash the concrete for you,
rise the river from the road.

You can run your hands over the valley,
trace the old riverbed.

Then I'll lay myself across it for you.
My hair becoming the watercress,

my bones the fish,
my blood the water,
my body the stony bed.

You can pile up
all of Wakefield's maps,
all of the deeds

and light them up.
Pāpā, we'll be ahi kā
once more.

JILLIAN SULLIVAN

Choosing

for Nick and Bex

How do you know
which hills and sky and water
will be your home,
the place where you long to return?

There's the unexpected beauty of light
in city structures on a lengthening night
beside the sea, the dark of furrowed loam,
an alabaster cottage, sheen of calm tide
through a wheelhouse window.

What of a river? Under the resilient arms
of willows, whatever the water says
over brown and shining stones,
you'll know if it's meant for you.

How do you choose
which rocks and trees and soil
will be your own?

Sometimes just by standing still
there with your feet on earth where you have landed
you'll feel the way two cogs within you
settle into unison,
power your heart, gain traction.

And when a bird lifts in the air above you,
something in your own heart
flings forward with a gust of joy,
the way a hawk soars, wingfeathers fanned,
riding the currents of desire
in a wide blue territory of sky.

APIRANA TAYLOR

my whenua

my whenua lies buried at the Karori dump
beneath asbestos dirty nappies and waste
my mother lies in Karori cemetery
they'll probably run a motorway through her bones
if i'd died as a child they'd have taken my heart without asking
pickled it and put it on display in their library
our urupa has been turned into a sewage-disposal unit
there are oxidation ponds
where once we caught the kokopu
don't be a stirrer I'm told
excuse me if i piss in your cup of tea

In view of the fact hospitals are meant to be places of healing,
I say, 'Long live Irihapeti Ramsden.'

ANTHONIE TONNON

Mataura Paper Mill

Mataura Paper Mill
words are lost on this waterfall
if you search out the photograph
you will feel the power it held

I went over the bridge
to bypass Gore
I saw a world
we just failed to fight for

I came down, I couldn't help myself
a neighbour drawn to watch the fire
but who am I to try to save this town
from a toxic ammonia cloud?

Mataura Paper Mill
hearts are lost in the detail
some protest
others not at all
what greater minds decide to mothball

and greater minds will always cross this bridge
to give their promise to the council
and greater minds always incorporate
so that no one can be held
responsible

Mataura Paper Mill
filled with dross from the smelter
all things flow in their natural course
to leave us with the liquidator

but I am here I couldn't help myself
a neighbour drawn to watch the fire
who were we, left to save this town
from a toxic, come on
from a toxic ammonia cloud?

Come on

BRIAN TURNER

Lament for the Taieri River

Everywhere I look
someone is trying to turn country I love,
dry lands studded with scattered craggy rock
and draped in bleached sparse grasses
tinted red, orange, yellow and white,
that wave urgently or desultorily
in the upland winds,
into country it was never meant to be.

Here strength and frailty go together,
and where water runs
it is lively, athletic, as water will.
There's laughter in it where the gravel's cleanest
and sunlight polishes and dazzles
when it silvers the flow.

But on the southern rise of the Maniototo plain,
on the terraced lift to the Rock and Pillar Range,
irrigation water flowers all day long,
twirls in lariat-like loops
over acres a lurid, anomalous green.
The Taieri River stinks from effluent,
is soured by pesticides and fertilisers.

The flow is sluggish, low;
shit and muck slosh into the river
where cattle mill and piss and break down banks
up to and beyond the Styx.

In the Patearoa pub a local farmer
waxed nostalgically about days of his youth
when, after coming home from school,

he 'could catch a fish on the threadline
within three casts. Something
needs to be done about the river,
it's pretty bad all right. But it's progress.
What do you do?'
'That's what we have a Regional Council for,'
I say. He knows that's true.

HONE TUWHARE

Pupurangi (Kauri Snail shell)

On the ground
that was densely
populated by
the Kauri,
I, Pupurangi once
thrived there also.

    The Kauri forest
    is gone. It is no more.
    But you may yet chance
    upon my spiral-conical
    house, intact, untenanted
    unshattered and unshat-upon
    by carefree, fourfooted cattle.

        I am no ordinary snail.
        I, Pupurangi—a long time
        fugitive from the Sea, have
        adapted to this sad environ
        of depletion, where the Kauri
        and me, are merely listed as endangered.

            In a time (when trees were plentiful)
            overactive fledglings that have fallen
            out of the nest, distract me by their
            feeble peeps, their helpless twitchings
            on the ground. I travel toward my target
            with great elan and swish. Moments like this
            do not find me gummily glum. There is a special
            lilt to the turbulence of air flowing over the dome
            of my house, as I lift my speed just a wee bit
            over familiar terrain. But now the Kauri and me
            in number—are diminished.

And now, if you hold me up to your ear
you may hear (painfully) the rising, agonized
shriek of the rip-saw biting—the thump-bump
                    of the hammer on the nail—driven.

TIM UPPERTON

Manawatū

The river twists like an eel
that twists within the twist

that is the river. The eel is a tube
that carries the river within it,

like the pipe that carries within it
what it pours into the river.

What pours all day into the river
becomes the river, as the child

who swims all day in the river
becomes a river-child,

and goes home in the evening
smelling of the river. The child

who swims in the river
dreams in the night of the river,

brown, flowing to the sea
with a burden it must disgorge,

and in that flow the flash
of the eel's upturned belly,

and the child, and everyone
the child has ever known,

faces upturned too and pale
in the moonlight, in the river,

so many faces, always more,
floating down the river

to where the river is going,
to where it widens, to the sea.

BRIAR WOOD

Whai

Me he ao e reti ana i te pō.
Don't call me paranoid!
Every year they're moving nearer—
this bay's being fished out

as orca come in close to feed.
See—an eagle ray in a wave
escaping like Darth Vader's cape,
Te Whai a Titipa in the night sky,

tell of stingray grey in the river
steely as a submarine flying saucer
with its blunt nose puggy stare
and curiously scapegrace flair

from Te Whai Wawewawe a Māui
or kites gliding in upcurrents—there—
while the electric ray's sharp rod
brings quick volts of recognition

emerging from flurries of mud
like the giant hand of a kaitiaki
plashing in the sand at low tide.
The barb's in the tail. Kia tūpato.

SUE WOOTTON

A Behoovement

'*Icebergs behoove the soul*'—**Elizabeth Bishop**, *'The Imaginary Iceberg'*

Seeks ceaselessly a spectrum space, one third afloat
and flashing in the squawking skyworld,
sculptured spectacle, sailing bright white

spectre-ship, two thirds submerged beneath
what splashes on the skin, in stately counterweight
to being awake: blue realm articulate in creaks and cracks
and booms. Prussian, midnight, cuttlefish, forget-
me-not. Behoven to its own and constant re-assemblage.
What is the soul to do, if the icebergs melt?

Notes

Introduction

1. John Felstiner, *Can Poetry Save the Earth?: A field guide to nature poems* (New Haven, CT: Yale University Press, 2009).
2. Gail Ingram, 'The poetry of getting back to living', *Corpus*, April 2019, par. 11.
3. The New Zealand Values Party, which was the precursor to the Green Party of Aotearoa New Zealand, was established early in this period, in 1972. Christine Dann, 'Experimental evolution down under: Thirty years of Green Party development in Australia and New Zealand', in *Green Parties in Transition: The end of grassroots democracy?*, eds E. Gene Frankland, et al., (Farnham, UK: Ashgate, 2008), p. 185.
4. British critic Jonathan Bate cites how Wordsworth 'connects his consciousness to the ecosystem' when recalling 'the blessed mood' that the memory of time spent in nature produces (from Wordsworth's celebrated poem 'Lines Written a Few Miles Above Tintern Abbey' (1798), cited in Jonathan Bate, *The Song of the Earth* (London: Picador, 2000), p. 147). Similarly, Elizabeth Bishop described the 'sweet / sensation of joy' passengers on a bus feel when a moose crosses the road in front of them, demonstrating the joy that comes when a connection to nature connects us to each other. Elizabeth Bishop, 'The Moose' (1946), cited in Bate, p. 202.
5. American critic J. Scott Bryson describes this as 'the interdependent nature of the world', *Ecopoetry: A critical introduction* (Salt Lake City: University of Utah Press, 2002), pp. 5–6.
6. David Mazel, 'American literary environmentalism as domestic orientalism', in *The Ecocriticism Reader: Landmarks in literary ecology*, eds Cheryll Glotfelty and Harold Fromm (Athens, GA: University of Georgia Press, 1996), p. 140.
7. Robert Hass, 'American ecopoetry: An introduction', in *The Ecopoetry Anthology*, eds Ann Fisher-Wirth and Laura-Gray Street, (San Antonio, TX: Trinity University Press, 2013), p. lii.
8. This conception broadens current definitions of ecology, such as 'the interrelations of all forms of plant and animal life with each other and their physical habitats', in which culture is absent. M.H. Abrams and Geoffrey G. Harpham, *A Glossary of Literary Terms*, 8th ed. (Boston: Thompson-Wadsworth, 2005), p. 71.
9. Ranginui Walker, *Ka Whawhai Tonu Matou: Struggle without end* (Auckland: Penguin, 1990), p. 11.
10. Some 400 traditional songs and chants from the iwi of Aotearoa can be found in *Ngā Moteatea* compiled by Sir Apirana Ngata (Ngāti Porou) and published from the 1920s onwards. This four-volume collection, with contributions from Pei Te Hurinui Jones (Ngāti Maniapoto), Tamati Reedy (Ngāti Porou) and Hirini Moko Mead (Ngāti Awa), translates, interprets and annotates the songs and chants. The song texts 'are evidence of the ancient and historical Māori tradition of composition, as well as the conventions and skills of the composer poets', say Jane McRae and Hēni Jacob in *Ngā Mōteatea: An introduction / He Kupu Arataki* (Auckland: Auckland University Press, 2011). In addition, the songs offer considerable information about Māori history and customary practices. The two most common kinds of songs collected in *Ngā Mōteatea* are waiata tangi, expressing grief or sadness, and waiata aroha, expressing love or longing. There are also oriori, songs sung to educate children, waiata whaiāipo that deal with desire, attraction or infatuation, and waiata whakautu or songs of reply that answer accusations, gossip or propositions. '[P]oetic and

highly allusive' recited songs or chants include haka (chants with posture dance), karakia (incantations), pātere, 'the often very long and strongly worded responses to accusations or mocking by others', and kaioraora, or cursing songs. There are also matakite, prophetic chants, and whakaaraara pā, chants recited by sentries as warnings, reassurance or to maintain their own alertness. Jane McRae and Hēni Jacob, *Ngā Mōteatea: An introduction / He kupu arataki* (Auckland: Auckland University Press, 2011), pp. 53–55.
11 Tim Flannery in *The Future Eaters: An ecological history of the Australasian lands and people* (Melbourne: Reed Books, 1994), p. 55.
12 Eric Pawson and Tom Brooking, *Making a New Land: Environmental histories of New Zealand* (Dunedin: Otago University Press, 2013), p. 89.
13 Ibid., p. 104.
14 Geoff Park, *Theatre Country: Essays on landscape and whenua* (Wellington: Victoria University Press, 2006), p. 183.
15 Ibid., p. 177.
16 Pawson and Brooking, *Making a New Land*, p. 26.
17 Rev. Maori Marsden, *The Woven Universe: Selected writings of Rev. Maori Marsden* (Ōtaki: Estate of Rev. Maori Marsden, 2003), p. 69.
18 Ibid., p. 111.
19 Mason Durie, *Ngā Tini Whetū: Navigating Māori futures* (Wellington: Huia, 2011), p. 263.
20 Philip Steer, 'Colonial ecologies: Guthrie-Smith's *Tutira* and writing in the settled environment', in *A History of New Zealand Literature*, ed. Mark Williams (Cambridge: Cambridge University Press, 2016), p. 85.
21 Arini Loader, 'Early Māori literature: The writing of Hakaraia Kiharo', in *A History of New Zealand Literature*, ed. Mark Williams (Cambridge: Cambridge University Press, 2016), p. 31.
22 Ibid.
23 Merata Kawharu, 'Environment as a marae locale', in *Māori and the Environment: Kaitiaki,* eds. Rachael Selby et al. (Wellington: Huia, 2010), pp. 222–24.
24 Patricia and Waiariki Grace, eds, *Earth, Sea, Sky: Images and Maori proverbs from the natural world of Aotearoa New Zealand* (Wellington: Huia, 2003).
25 Ibid., p. 14.
26 Ibid., p. 72.
27 Alex Preminger and T.V.F. Brogan, eds, *The New Princeton Encyclopedia of Poetry and Poetics* (Princeton: Princeton University Press, 1993), p. 166.
28 Witi Ihimaera and D.S. Long (eds), *Into the World of Light: An anthology of Maori writing* (Auckland: Heinemann, 1982), p. 2.
29 Te Kapunga Matemoana (Koro) Dewes, *Māori Literature: A tentative framework for study* (1974), quoted in Kelly Lambert, 'Calling the taniwha: Mana wahine Maori and the poetry of Roma Potiki' (Master of Arts thesis, Victoria University of Wellington, 2006), p. 80.
30 Miriama Evans, *The Penguin Book of Contemporary New Zealand Poetry: Ngā kopu tītohu o Aotearoa* (Auckland: Penguin, 1989), p. 19.
31 Anne Salmond, *Tears of Rangi: Experiments across worlds* (Auckland: Auckland University Press, 2017), p. 313.
32 Tina Ngata, 'Wai Māori', in *Mountains to the Sea: Solving New Zealand's freshwater crisis,* ed. Mike Joy (Wellington: Bridget Williams Books, 2018), p. 28.
33 Salmond, *Tears of Rangi*, p. 293.
34 Dan Cheater, 'I am the River and the River is me: Legal personhood and emerging rights of nature', *West Coast Environmental Law,* March 22, 2018.
35 Salmond, *Tears of Rangi*, p. 313.
36 Ibid., p. 410.
37 Melissa Kennedy, 'The Māori Renaissance from 1972', in *A History of New Zealand Literature*, ed. Mark Williams (Cambridge: Cambridge University Press, 2016), p. 277.
38 Ibid.
39 John Huria, 'Review of *Deep River Talk: Collected Poems* by Hone Tuwhare', *Landfall,* no. 186, 1993, p. 335.
40 Ibid.
41 Jane McRae, *Māori Oral Tradition: He kōrero nō te ao tawhiti* (Auckland: Auckland University Press, 2017), p. 25.
42 Bernard Gadd, 'Hone Tuwhare in his poetry', *Landfall,* no. 149, 1984, p. 84.

43 Kennedy, 'The Māori Renaissance from 1972', p. 281.
44 Witi Ihimaera and Tina Makereti, *Black Marks on the White Page* (Auckland: RHNZ Vintage, 2017), p. 8.
45 Ibid., p. 13.
46 Ibid., p. 10.
47 Elizabeth M. DeLoughrey, 'Introduction: A postcolonial environmental humanities', in *Global Ecologies and the Environmental Humanities: Postcolonial approaches*, eds E. DeLoughrey et al., (London: Routledge, 2015), pp. 1, 5.
48 Jacob Edmond, 'Against global literary studies', *New Global Studies* 15, 2–3, 2021, p. 214.

The early years

1 Anahera Gildea, 'Kōiwi pāmamao: The distance in our bones', *The Pantograph Punch*, 2 April 2008, https://pantograph-punch.com/posts/bones
2 Merata Kawharu, *Tāhuhu Kōrero: The sayings of Taitokerau* (Auckland: Auckland University Press, 2008), p. 139.
3 Ibid., p. 140.
4 Ibid.
5 Translated by Pei Te Hurinui Jones in *Ngā Mōteatea: He maramara rere nō ngā waka maha / The Songs: scattered pieces from many canoe areas* (1928–88) collected by Sir Apirana Ngata, pp. 398–99.
6 Jane McRae and Hēni Jacob, *Ngā Mōteatea: An introduction / He kupu arataki* (Auckland: Auckland University Press, 2011), p. 59. The kōkako also invokes the story of how this bird brought water for a thirsty Māui, who in gratitude stretched the bird's legs, giving it a longer stride.
7 Ibid.
8 Shared by Te Ahukaramū Charles Royal in his report 'Marine disposal of wastes; a Māori view' (1989), and quoted by A.M. Jackson, N. Mita and H. Hakopa in their report, 'Hui-te-ana-nui: Understanding kaitiakitanga in our marine environment' (July 2017).
9 Ibid., p. 97: Jackson et al. quotation from the Waitangi Tribunal report, 'Ko Aotearoa tēnei: A report into claims concerning New Zealand law and policy affecting Māori culture and identity', (2011), p. 105.
10 Published in *Te Mareikura* (1 November 1911) and translated by folk song anthologist John Archer on his website 'New Zealand Waiata Matakite', www.folksong.org.nz
11 Te Kooti's waiata is included in the manuscript of waiata complied by Hamiora Aparoa and donated to the University of Auckland Library by Sir Monita Delamere, and in the compilation of transcripts of the same waiata made by Robert Biddle. Biddle's material is held by his son, the secretary of the Haahi Ringatu. Judith Binney, *Redemption Songs: A life of Te Kooti Arikirangi Te Turuki* (Wellington: Bridget Williams Books, 1995), p. 561.
12 In 1972, the mātuhituhi had the unfortunate distinction of being one of the last Aotearoa New Zealand birds to become extinct.
13 Quoted in Sarah Shieff, *Talking Music: Conversations with New Zealand musicians* (Auckland: Auckland University Press, 2002), p. 145.
14 Margaret Orbell, *Maori Poetry: An introductory anthology* (Auckland: Heinemann Educational, 1978), p. 84.
15 Ibid.
16 Translated by John Archer on his website www.folksong.org.nz

The middle years

1 Herbert Guthrie-Smith, *Tutira: The story of a New Zealand sheep station* (Edinburgh: Blackwood, 1953), p. xxiii.
2 Herewini Easton, 'He Kōrero Paki nō Tawhiti mai: Narratives from distant past' (Master of Creative Writing thesis, Auckland University of Technology, 2020), p. 4.
3 Ibid., p. 29.
4 Rachael Te Āwhina Ka'ai-Mahuta, 'He kupu tuku iho mō tēnei reanga: A critical analysis of Waiata and Haka as commentaries and archives of Māori political history' (PhD thesis, Auckland University of Technology, 2010), p. 221.

About the Poets

FLEUR ADCOCK CNZM OBE is a poet and editor. Her poetry frequently investigates questions of identity and belonging, and the natural worlds of England and New Zealand. In 1996, she was awarded an Order of the British Empire for her contribution to New Zealand literature, and in 2008 she was made a Companion of the New Zealand Order of Merit for services to literature.

JENNY ARGANTE was a literary midwife, helping to deliver other writers' books for two decades. Invigorated by what she learned, she is now returning to her own poetry and prose and working on follow-ups to *After the Act* and *Working in the Cracks Between* from Oceanbooks.

BRIDGET AUCHMUTY lives in a yurt in Central Otago, a rich seam for writing about landscape and environment. Her fiction and poetry have appeared in various journals, including *JAAM*, *Mslexia*, *Meniscus* and *takahē*. She holds a Master of Creative Writing from Massey University, and her first poetry collection, *Unmooring*, was published in 2020.

TUSIATA AVIA has published five books of poetry. *Wild Dogs Under My Skirt* and *The Savage Coloniser Book* are also successful stage shows. *The Savage Coloniser Book* won the Mary and Peter Biggs Award for Poetry at the 2021 Ockham New Zealand Book Awards.

HINEMOANA BAKER (Ngāti Raukawa, Ngāti Toa, Te Āti Awa, Ngāi Tahu, Germany, England) is the author of four collections of poetry. A bilingual edition of her latest, *Funkhaus* (Te Herenga Waka University Press, 2020), was released in 2023 by Voland & Quist in Berlin (German and English, tr. Ulrike Almut Sandig).

ALEXANDER BATHGATE (1845–1930) migrated from Edinburgh to Ōtepoti Dunedin in 1863. He was a lawyer, company director, newspaper columnist, author and founding member of Aotearoa New Zealand's first conservation group, the Dunedin and Suburban Reserves Conservation Society.

BLANCHE [EDITH] BAUGHAN (1870–1958) was a poet, writer and penal reformer. Born in London, she was one of the first women to graduate from Royal Holloway College. She was active in social work in the slums of east London and the English suffrage movement before travelling extensively. In 1902, she settled in Banks Peninsula, where she wrote poetry and travel essays.

JAMES K. BAXTER (1926–1972) was a poet, playwright and activist for the preservation of Māori culture. He was regarded as the preeminent writer of his generation. In the 1950s, he was associated with the Wellington group of poets, taking an increasingly strong stance towards what he saw as the spiritual deadness of modern middle-class life. In the late 1960s, he established a commune for those alienated from city life on Māori land at Hiruhārama, Jerusalem, on the Whanganui River.

AIRINI BEAUTRAIS is a poet, writer and educator who lives in Whanganui. In 2021, she won the Jann Medlicott Acorn Prize for Fiction for *Bug Week*. Her most recent work is a collection of essays, *The Beautiful Afternoon* (Te Herenga Waka University Press, 2023).

URSULA BETHELL (1874–1945) was born in England and came to Aotearoa New Zealand as a baby. Her family settled in Rangiora and the Ashley River and Mt Grey were significant landmarks in her life and poetry. She completed her education in Oxford and then Switzerland. She was a lifelong Anglican and her faith informed her poetry and her social work. She returned to Ōtautahi Christchurch in 1924, settling at Rise Cottage on the Cashmere Hills, where her garden was the subject of many of her poems.

PETER BLAND is a poet, writer, actor and theatre director. He was born in North Yorkshire and emigrated to New Zealand in 1954 where he graduated from Victoria University. He was a member of the Wellington group of poets and his poetry criticised this country's culture of conformity. He worked as a radio producer at the NZBC, and was co-founder and artistic director of Downstage Theatre 1964–68. He lives in Tāmaki Makaurau Auckland and has recorded readings of his poems on Facebook.

ARAPERA HINEIRA KAA BLANK (Ngāti Porou, Ngāti Kahungunu, Rongowhakaata, Te Aitanga a Māhaki) (1932–2002) was a poet, short story writer and teacher. She graduated from Auckland University and taught for 25 years. She was one of this country's first bilingual poets, and one of the first Māori writers to be published in English. Her work, which focuses on Māori life and the life of women, engages Māori and Pākehā world views.

CHARLES BRASCH (1909–1973) was a poet, literary editor and arts patron. He was born in Ōtepoti Dunedin and studied at Oxford University. His exploration of European settlement in New Zealand was a theme shared by Pākehā writers of his generation. In 1946, he founded *Landfall* and during 20 years of editing the literary journal he had a significant impact on the development of this country's literary and artistic culture in English.

BRENDA BURKE was born in Canada and has lived in Pōneke Wellington.

ALISTAIR TE ARIKI CAMPBELL ONZM (1925–2009) was a poet, playwright, novelist and editor. He was one of this country's most distinctive poetic voices from the 1950s to the 2000s. Shaped by his Rarotongan childhood, exile to New Zealand and a transformative return to Polynesia in middle age, his work traces shifting relations between English, Māori and Pacific peoples' traditions. In 2005, he was awarded the Order of New Zealand Merit for literary achievement in poetry.

JACQ CARTER (Ngāi Te Rangi, Ngāti Awa, Waitaha and Ngāti Pākehā, with links to Ngāi Tai, Ngāti Maru and Ngāpuhi Nui Tonu) has been published in various anthologies, including Robert Sullivan, Albert Wendt and Reina Whaitiri's *Whetū Moana*, its sequel *Mauri Ola*, and all editions of Anton Blank's *Ora Nui*.

ALLEN CURNOW (1911–2001) has been described as one of the greatest twentieth-century poets writing in English. For 70 years his poetry was always on the move – from his early approaches to New Zealand identity and myth to the scrupulous attention to memory, mortality and reflection of the later poems. He played a major role in New Zealand's cultural life as a critic, anthologist and playwright, and phrases from his poems have entered the national discourse.

JONATHAN CWEORTH is an Ōtepoti Dunedin poet and playwright. His poems have been published in journals and anthologies, including *Poetry Aotearoa Yearbook*, the *Otago Daily Times* and *Manifesto Aotearoa*. A number of his scripts have been performed in local arts festivals. He loves working in writing groups.

RUTH DALLAS (1919–2008) is one of Aotearoa New Zealand's most distinguished poets. She was born in Invercargill and lived for many years in Ōtepoti Dunedin. Her poetry, imbued with the landscapes of the southern South Island, reflects the rhythms of the natural world and the human transience within it. She published many poetry collections and children's books and won several literary awards.

HARRY DANSEY MBE (Ngāti Tūwharetoa, Te Arawa, Ngāti Raukawa) (1920–1979) was a journalist, cartoonist, writer, broadcaster, local politician and race relations conciliator. From 1961–70, he was the *Auckland Star* writer on Māori and Pacific Island affairs. In 1974, he was made a Member of the British Empire for services to journalism.

LYNN DAVIDSON's poem 'Speaking to the Otter' appears in her latest poetry collection *Islander* (Shearsman Books and Te Herenga Waka University Press, 2019) and *Project Boast* (Triarchy Press, 2018). Her memoir *Do You Still Have Time for Chaos?* was published by Te Herenga Waka University Press in 2024.

LAURIS EDMOND OBE (1924–2000) is celebrated as one of the finest poets of late twentieth-century Aotearoa New Zealand. She published her first collection when she was 51 and quickly garnered recognition for her uniquely mature yet vibrant voice. She went on to write a novel, short stories and stage dramas, 11 volumes of poetry and a three-volume autobiography. Her *Selected Poems* won the Commonwealth Poetry Prize in 1985, she was made an OBE for services to poetry and literature in 1986, and in 1999 she received a lifetime achievement award at the Montana New Zealand Book Awards.

DAVID EGGLETON is a performance poet, writer, editor, freelance journalist and critic. Of Rotuman, Tongan and Palagi ancestry, he was born in Tāmaki Makaurau Auckland and grew up in Fiji and Aotearoa New Zealand. Described as a beatnik pop poet, he began reciting his poetry at rock music gigs in the early 1980s – he was voted *TimeOut* Street Poet of the year in 1985 – and he remains interested in presenting poetry in a variety of media contexts. His frequently anti-establishment poems, in 18 collections, subvert and mock the language of media, politics and corporates. He was New Zealand Poet Laureate 2019–22.

TE MĀREIKURA HORI ENOKA (1877–1946) was a lyricist whose manuscripts include more than 4000 pao and hundreds of waiata. He was born in Waikato and died in Ohakune. His work was influenced by Rangitikei prophetess Mere Rikiriki, and melded Māori spirituality with the Catholic faith.

RANGI FAITH (Kāi Tahu, Ngāti Kahungunu, English, Scottish) was born in Timaru to English and Māori parents and brought up in South Canterbury. He is retired from teaching and is currently living in Rangiora. His work explores European and Māori history and welcomes the resurgence of te reo Māori in Aotearoa New Zealand.

FIONA FARRELL ONZM publishes poetry, plays, fiction and non-fiction. She has received numerous awards, including a New Zealand Book Award for Fiction, Creative New Zealand's premier Michael King Fellowship and the Katherine Mansfield Menton Fellowship. In 2007, she was awarded the Prime Minister's Award for Fiction and in 2012 the Order of New Zealand Merit for Services to Literature.

SUE FITCHETT is a conservationist, volunteer firefighter, Waiheke Islander, and the author of *Palaver Lava Queen* (Auckland University Press, 2004) and *On the Wing* (Steele Roberts, 2014). She is the co-author or editor of several poetry books and anthologies. She was the Louis Johnson Bursar 2001–2002.

ALISON GLENNY is the author of *The Farewell Tourist*, which won the 2017 Kathleen Grattan Poetry Award and was published by Otago University Press in 2018, and *Bird Collector* (Compound Press, 2022). She lives on the Kāpiti Coast.

TRICIA GLENSOR lives on Pōneke Wellington's south coast and works as a freelance writer and editor. Her novel, *Telling Lies*, was published by HarperCollins in 2012.

DENIS GLOVER DSC (1912–1980) was a poet, journalist, printer, typographer, publisher and naval officer. He was born in Ōtepoti Dunedin and graduated from Canterbury College, where he became an assistant lecturer. In 1936, he founded the Caxton Press, which published *Landfall* and the work of many established names in New Zealand literature in English, including himself. He published many collections of poetry and became a legend in his lifetime for his lyrical talent and irreverence. His 'The Magpies', is this country's most widely anthologised poem.

PATRICIA GRACE (Ngāti Toa, Ngāti Raukawa, Te Āti Awa) is a key figure in the emergence of Māori fiction in English since the 1970s and has made a significant contribution to contemporary New Zealand literature. Exploring themes such as loss, isolation and family, she portrays a variety of Māori people and ways of life and is notable for her versatile narrative and descriptive techniques. Her best-known novel, *Potiki* (1986), won the New Zealand Book Award for Fiction and has been translated into several languages.

KEREHI WAIARIKI GRACE (Ngāti Porou, Te Whānau a Apanui) (1936–2013) was a Māori culture expert with a background in education. He was the author of several children's books and wrote *Earth, Sea and Sky* with Patricia Grace.

JORDAN HAMEL is an Aotearoa New Zealand writer and performer. His debut collection *Everyone is Everyone Except You* was published by Dead Bird Books in 2022 and will be published in the UK in 2024. He is also the co-editor of an anthology of New Zealand climate change poetry, *No Other Place to Stand* (Auckland University Press, 2022). His recent work can be found or is forthcoming in *POETRY*, *Sonora Review*, *Gulf Coast* and *Ōrongohau | Best New Zealand Poems*.

HOROMONA HAPAI (Ngāti Porou) d. 1917.

RANGI TANIRA HARRISON (Tangata marae, Okauia, Matamata. Ngāti Tangata, Ngāti Hinerangi, Ngāti Raukawa, Ngāti Kotimana) (1924–1981) composed the waiata 'Waikato te Awa' in 1961–62 while working in Mangakino. It was published in *Te Ao Hou* (June 1962). 'Waikato te Awa' is in the tempo of pātere, a chant that flows.

DINAH HAWKEN's ninth poetry collection, *Sea-light*, was published by Te Herenga Waka University Press in 2021. Her first book won the 1987 Commonwealth Poetry Prize for Best First Book, and four subsequent collections have been shortlisted for the New Zealand Book Awards. She was born in Hawera in 1943, has lived in Pōneke Wellington and New York and now lives in Paekakariki.

REBECCA HAWKES is a painter-poet raised among ruminants on a sheep and beef farm near Methven. Her book *MEAT LOVERS* was unleashed by Auckland University Press in 2022, winning Best First International Collection in the UK Laureate's Laurel Prize and shortlisted for a Lambda Literary Award. She edits hot-blooded journal *Sweet Mammalian* and co-edited the climate poetics anthology *No Other Place to Stand*. Rebecca's chapbooks 'Softcore Coldsores' and 'Hardcore Pastorals' can be found in *AUP New Poets 5* and *Cordite*.

HELEN HEATH's debut collection, *Graft*, won the NZSA Jessie Mackay Best First Book of Poetry Prize in 2013 and was the first book of fiction or poetry to be shortlisted for the Royal Society of New Zealand Science Book Prize. Her second collection, *Are Friends Electric?*, won the 2019 Mary and Peter Biggs Award for Poetry at the Ockham New Zealand Book Awards.

NOLEEN HOOD lived on the West Coast in the 1970s and retired to Ōtautahi Christchurch in 1989. Her poetry collection *Trim Pork and Waterfalls* (1991) was published by Ledingham Books, Blaketown.

PETER HOOPER [HEDLEY COLWILL] (1919–1991) was a teacher, writer, bookseller and conservationist. He was born in London and emigrated to Aotearoa New Zealand at the age of four. He ran Walden Books, named for *Walden* (1854) by American transcendentalist writer Henry David Thoreau, a book which influenced his views on nature and living. He was awarded the 1990 New Zealand Commemoration Medal.

JOHN HOWELL lives in Ngaio. His first book of poems was *Homeless* (Mākaro Press, 2017). He is a retired Minister and has written two books of prayers. He was OUSA President in 1972.

KERI HULME (Ngāi Tahu, Kāi Te Ruahikihiki, Orcadian Scots, Lancashire English) (1947–2021) was a novelist, poet and short story writer. Keri was born in Ōtautahi Christchurch and lived for almost 40 years in Ōkārito in south Westland. Keri's debut novel, *The Bone People* (1984), won the Booker Prize in 1985. Her writing explores themes of isolation, identity, spirituality, mythology and tikanga Māori. In 1990 Keri was awarded the New Zealand Commemoration Medal.

SAM HUNT CNZM QSM is a poet. He became widely known in the 1970s and 1980s as a performer of his own work, and the work of others, especially James K. Baxter, who was his friend and influencer. He performed in a variety of venues throughout the country, often in the company of his dog, Minstrel. A self-described 'itinerant minstrel', he became a nationally recognised artist. In 1985 he was awarded the Queen's Service Medal for community service, and in 2010 was appointed a Companion of the New Zealand Order of Merit for services to poetry.

FRANCIS HUTCHINSON (c. 1868–1940) was a farmer, writer and naturalist. He bred stud Romney sheep at Rissington, Hawke's Bay and edited *The East Coast Naturalist* 1899–1904.

GAIL INGRAM is an award-winning writer from Ōtautahi Christchurch and the author of the poetry collections *Some Bird* (Sudden Valley Press, 2023) and *Contents Under Pressure* (Pūkeko Publications, 2019). Her work has appeared widely across Aotearoa and in Australia, Africa, the UK and the USA. She is an editor for *a fine line* and *Flash Frontier*.

KEVIN IRELAND OBE (1933–2023) was a poet, short story writer, novelist and librettist. As a young writer, he lived in Frank Sargeson's bach. He was co-founder of the literary magazine *Mate*. From 1959, he lived overseas, mainly in England, where he worked as a printer's reader at *The Times* for 25 years, consistently identifying as a New Zealand poet. He published 20 collections of poetry, six novels, numerous short stories and two memoirs. In 1992, he was appointed an Officer of the Order of the British Empire for services to literature.

ANNA JACKSON is a poet, fiction and non-fiction writer and academic. She was born in Tāmaki Makaurau Auckland, studied English at Auckland and Oxford universities, and is associate professor of English literature at Victoria University Te Herenga Waka. She has published books of poetry, fiction and critical writing about poetry.

ASH DAVIDA JANE is a poet and editor from Te Whanganui-a-Tara. Her second book, *How to Live With Mammals* (Te Herenga Waka University Press, 2021), won second prize in the Laurel Prize, an international award for environmental poetry. She is a publisher at Tender Press and reviews books for Radio New Zealand.

TIM JONES was awarded the NZSA Peter & Dianne Beatson Fellowship in 2022. His recent books include the poetry collection *New Sea Land* (Mākaro Press, 2016) and the climate fiction novella *Where We Land* (The Cuba Press, 2019). His new climate fiction novel *Emergency Weather* was published by The Cuba Press in 2023.

ANNE KENNEDY's recent books are *The Sea Walks into a Wall*, *The Ice Shelf*, and, as editor, *Remember Me: Poems to learn by heart from Aotearoa New Zealand*. Her awards include the Prime Minister's Award for Literary Achievement and the Montana New Zealand Book Award for Poetry.

ERIK KENNEDY is the author of *Another Beautiful Day Indoors* (2022) and *There's No Place Like the*

ABOUT THE POETS

Internet in Springtime (2018), both with Te Herenga Waka University Press, and co-editor of *No Other Place to Stand* (Auckland University Press, 2022), an anthology of climate change poetry from Aotearoa New Zealand and the Pacific.

MEGAN KITCHING is an Ōtepoti Dunedin poet. Her debut collection, *At the Point of Seeing* (Otago University Press, 2023), won the Jessie Mackay Prize for Poetry at the 2024 Ockham New Zealand Book Awards. In 2021, she was the inaugural Caselberg Trust Elizabeth Brooke-Carr Emerging Writer Resident.

LEICESTER KYLE (1937– 2006) was a poet, priest, environmentalist and scientist. Born in Ōtautahi Christchurch, he was a minister in the Anglican Church before retiring to Tāmaki Makaurau Auckland in 1995. In 1998, he moved to Millerton on the West Coast of the South Island, where he wrote ecological poems in protest against proposed strip-mining of the Millerton plateau. In 2014, literary executors Jack Ross and David Howard posthumously published his later work.

ARIHIA LATHAM (Kāi Tahu, Kāti Māmoe, Waitaha) is a writer, creative, and rongoā practitioner. Her poetry collection *Birdspeak* is published by Anahera Press (2023), and her short stories, essays and poetry have been widely published and anthologised. She lives with her whānau in Pōneke Wellington.

HELEN LEHNDORF's latest book, *A Forager's Life*, is a creative non-fiction nature memoir. It made the top ten list for New Zealand non-fiction for several weeks in 2023. She is the author of the poetry collection *The Comforter*, which made *The Listener*'s 100 Best Books list in 2011 and *Write to the Centre*, a book about the benefits of keeping a journal.

TOI TE RITO MAIHI (Ngāti Kahungunu, Ngāpuhi) (1937–2022) was a poet, writer, teacher, illustrator and weaver. She was a foundation member of Aotearoa Moananui-a-Kiwa Weavers (now known as Te Roopu Raranga Whatu o Aotearoa). Her written work traced the development of Māori fabric-making techniques, and her artistic work presented the aspirations of Waitangi Tribunal claimant groups in visual form.

BILL MANHIRE CNZM is a poet, editor, short story writer and teacher. He has published many collections of poetry and edited several major anthologies of poems and short stories. In 1997–98, he was Aotearoa New Zealand's first Poet Laureate, and in 2005, he was appointed a Companion of the New Zealand Order of Merit for services to literature.

SELINA TUSITALA MARSH is a poet, academic and performer. She was born in Tāmaki Makaurau and was the first Pacific Islander to complete a doctorate in English at Auckland University. In 2012, she represented Tuvalu at the Cultural Olympiad Poetry Parnassus Festival in London, and in 2016 was named the official Commonwealth poet. Her scholarly work focuses on Māori and Pacific literature and culture, and she edits and maintains *Pasifika Poetry*, a website devoted to preserving and curating the work of Pacific poets.

CAROLYN McCURDIE is an Ōtepoti Dunedin writer. Her first collection of poetry *Bones in the Octagon* was published by Makaro Press in 2015. She is working towards a second collection.

DONALD McDONALD (1911–42) was a farmer, soldier and poet. He joined the army in 1940 and served in Fiji and North Africa, where he was wounded at the 1941 battle of Sidi Reszegh in Libya. In hospital, he wrote the verses, 'Sidi Reszegh', which he posted home to his mother. He was taken prisoner at El Alamein and killed in an Italian vessel torpedoed in the Mediterranean. His mother found many of his poems, which were collected and published by his former school.

MARIA McMILLAN was born in Ōtautahi Christchurch and now lives on the Kāpiti Coast. Her most recent book is *The Ski Flier* (Te Herenga Waka University Press, 2017). 'How they came to privatise the night', was written in response to the repeated attempts by private corporations to take over the world's freshwater supply.

CILLA McQUEEN lives in Motupōhue Bluff, at the southern tip of Te Waipounamu South Island. A poet and artist, she has published 17 collections and a CD of her poetry. She was New Zealand Poet Laureate 2009–11 and in 2010 received the Prime Minister's Award for Poetry.

VAN MEI (they/them) is an artist and writer of Hokkien Chinese and Pākehā descent. They are the current Kaitohu of The Pantograph Punch, an Aotearoa-based online arts and culture platform. Prior to this, they were director of Enjoy Contemporary Art Space, and before that, they were playing Stardew Valley while lying on the couch.

HIRINI MELBOURNE ONZM (Ngāi Tūhoe, Ngāti Kahungunu) (1949–2003) was a composer, singer, university lecturer, poet and author. He was influential in the revival of te reo Māori music and culture. With ethnomusicologist and performer Richard Nunns, he performed and made recordings that contributed to the ongoing ngā taonga pūoro revival. In 2003, he was appointed an Officer of the New Zealand Order of Merit for services to Māori language, music and culture.

TRIXIE TE ARAMA MENZIES (Ngāti Hei, Ngāti Whanaunga, Ngāti Maru) (1936–2017) was a poet and former teacher. She was one of the first Māori women poets published in English. Her first of four poetry collections was published in 1986. She was a founding member of Waiata Koa, the Māori Women's Artists and Writers Collective formed at the 1986 Karanga Karanga exhibition at City Gallery Wellington, which was the first public museum show of collaborative works by Māori women artists.

KARLO MILA MNZM is a poet, writer and scholar. Her poetry and scholarship focus on the personal and political realities of Pasifika identity. She is director of Mana Moana Leadership New Zealand, which aims to vitalise and prioritise Pasifika ancestral knowledge in contemporary contexts. In 2012, she represented Tonga at the Cultural Olympiad Poetry Parnassus Festival in London, and in 2019, she was appointed a Member of the New Zealand Order of Merit for services to the Pacific community and as a poet.

BARRY MITCALFE (1930–1986) was a poet, editor and peace activist. He was born in Pōneke Wellington and taught Māori and Polynesian studies. In the 1960s and 1970s, he was a leader of this country's anti-Vietnam War and anti-nuclear protest movements. In 1977, he was awarded the Katherine Mansfield Fellowship in Menton. He lived on Coromandel Peninsula and founded the Coromandel Press.

HARVEY MOLLOY lives in Pōneke Wellington. He is the author of three books of poetry: *Night Music* (2018), *Udon by The Remarkables* (2016) and *Moonshot* (2008). His poetry and flash fiction have been published in anthologies, including *Ōrongohau | Best New Zealand Poems*, *Essential New Zealand Poems* and *Bonsai: Best small stories from Aotearoa New Zealand*.

ELIZABETH MORTON is a Tāmaki Makaurau Auckland writer with three poetry collections, the latest being *Naming the Beasts* (Otago University Press, 2022). She is a part-time academic in the domains of neuroscience and mental health. Her yarns and poems have appeared in journals and anthologies locally and in Canada, Australia, the USA, the UK, Ireland, Italy and Jamaica.

ALAN MULGAN OBE (1881–1962) was a journalist, writer and broadcaster. He was chief leader writer at the *Auckland Star* 1916–35, and foundation lecturer in journalism at Auckland University College 1924–45. His books and poems reflect the transition between Victorian and nationalist ideals. In 1947, he was appointed Officer of the Order of the British Empire for services to literature, journalism and broadcasting.

SAANA MURRAY CNZM QSM (Ngāti Kurī) (1925–2011) was a teacher, master weaver, writer and Tiriti o Waitangi activist. She composed waiata and poetry in English and te reo Māori that helped to catalyse the 1975 Māori Hīkoi. She was one of six original claimants of Wai 262, which sought to restore tino rangitiratanga through Māori involvement in decision-making regarding flora, fauna and taonga. In 2009, she was made a Companion of the New Zealand Order of Merit for contributions to Māori.

EMMA NEALE has published six novels, six poetry collections, a collection of short stories, and she has edited several anthologies. She was editor of *Landfall* from 2018 to 2021 and received the Lauris Edmond Memorial Award for Distinguished Contribution to New Zealand Poetry in 2020.

JOHN NEWTON is a poet, academic, critic, cultural historian and musician. He grew up on a sheep farm at Port Underwood in the Marlborough Sounds. His three collections of poetry reflect on his provincial backblocks upbringing and real and imagined versions of this country's landscape. He has written a verse novel, a history of James K. Baxter and his time spent establishing a commune, the first instalment of an Aotearoa New Zealand literature trilogy, and a book about the life and works of sculptor Llew Summers.

NAOMI O'CONNOR is a writer and editor living in Pōneke Wellington. She grew up in a beach suburb on the edge of Ōtautahi Christchurch, looking across to the Kaikōura mountains. If she goes to the coast on a good day in Te Whanganui-a-Tara she can see the other end of the same ranges.

CHRIS ORSMAN was born in Lower Hutt and lives in Pōneke Wellington. He has published two collections of poetry, and four chapbooks published by his own poetry label, Pemmican Press. In 1998, he was one of the first artists to be awarded the Antarctica Arts Fellowship.

EVELYN PATUAWA-NATHAN (Te Roroa, Ngāti Whatua, Te Rarawa, Ngāpuhi, Ngāti Torehina, Ngāti Hau, Ngāti Maniapoto) (1933–2019) was a poet and painter. She was one of the first Māori women poets published in English with her 1979 collection *Opening Doors*, which was published in Fiji. Her poetry and art depicts Māori customs and their loss, and ecological degradation of the Kaihu River and Tutamoe Range of her childhood.

MERIMERI PENFOLD CNZM (Ngāti Kurī) (1924–2014) was a poet, teacher and translator. She was born at Te Hāpua, Northland and educated at Queen Victoria School, the University Coaching College where her Latin teacher was R.A.K. Mason, and Auckland Girls' Grammar School. She was a senior lecturer in Māori language at Auckland University where she also taught a course in Māori weaving and plaitwork. In 2001 she was appointed a Companion of the New Zealand Order of Merit for services to Māori.

VERNICE WINEERA PERE (Ngāti Toa Rangitira, Ngāti Raukawa) is a poet, academic and descendent of Te Rauparaha. She was born in Pōneke Wellington and grew up at Takapuwahia Pa, Porirua and now lives in La`ie, Hawai`i, while maintaining her strong whakapapa, turangawaewae and whānau connections with Aotearoa. She was the first Māori woman to achieve a published voice in poetry in English when her debut book of poems was published in 1978.

KIRI PIAHANA-WONG is a poet and editor and the publisher at Anahera Press. Her most recent publication is *Te Awa o Kupu*, an anthology of contemporary Māori writing co-edited with Vaughan Rapatahana (Penguin Random House, 2023). Kiri lives in Whanganui with her family.

ROBYN MAREE PICKENS is an art writer and poet who lives with her kitten Pippin in Ōtepoti Dunedin. Her first poetry collection, *tung*, was published by Otago University Press in 2023. She received her doctorate in reparative ecopoetics from the University of Otago in 2022.

NINA MINGYA POWLES is a poet and writer based in the UK. Her poetry collection *Magnolia 木蘭* was published in the UK, NZ and US, and shortlisted for the Ockham New Zealand Book Awards. She is also the author of several poetry pamphlets and zines, and is on the *Starling* editorial committee. Her essay collection *Small Bodies of Water* was published in 2021.

ROMA POTIKI (Te Rarawa, Te Aupouri, Ngāti Rangitihi) is a poet, playwright, visual artist, curator, theatre actor and director, and commentator on Māori theatre. She was born in Lower Hutt, where her art is in the permanent collection of the Dowse Art Museum. She was involved in contemporary Māori theatre in its formative years and, in 1979, helped to form the theatre company Maranga Mai, which toured with a production about Māori activism. In 1989, she founded He Ara Hou Theatre Māori Incorporated.

PUA-RORO (Ngāti Toa) was a chief of Te Totara pa, a prominent point south of Kawhia Heads, in about 1670–75. He was a composer of mōteatea.

VAUGHAN RAPATAHANA (Te Ātiawa) commutes between homes in Hong Kong, the Philippines and Aotearoa New Zealand. He is widely published across several genres in both his main languages, te reo Māori and English, and his work has been translated into Bahasa Malaysia, Italian, French, Mandarin, Romanian and Spanish. He is the author, editor or co-editor of more than 40 books.

RICHARD REEVE is the author of six poetry collections: *Dialectic of Mud* (Auckland University Press, 2001), *The Life and the Dark* (Auckland University Press, 2004), *In Continents* (Auckland University Press, 2008), *The Among* (Maungatua Press, 2008), *Generation Kitchen* (Otago University Press, 2015), and *Horse and Sheep* (Maungatua Press, 2019). His latest poetry collection, *About Now*, was published in 2024.

WILLIAM PEMBER REEVES (1857–1932) was a politician, historian, journalist, banker and poet. Born in Lyttelton to wealthy settlers, he became editor of the *Canterbury Times* in 1885 and a Member of Parliament from 1887–96. In 1896 he was made New Zealand agent-general in London. He wrote poetry, short stories, and on New Zealand history and social reform.

HARRY RICKETTS is a poet, editor, reviewer, author, academic and cricket writer. He was born in London, studied English at Oxford University, and lectured in Hong Kong and Leicester before joining the English Department at Victoria University in 1981. He has published poetry and non-fiction, including a study of WWI poets, and is the co-author with Paula Green of *99 Ways into New Zealand Poetry* (2010).

TE KAHU ROLLESTON He uri a Te Kahu i ngā ngaru e papaki tū ana ki Mauao. Tauranga moana, Tauranga tāngata eeeee.

SAM SAMPSON was born in Tāmaki Makaurau Auckland and raised in South Titirangi, next to Little Muddy Creek (Waikumete), where he still lives. Sampson's most recent collection, *Un Coup de Dés N'abolira le Hasard (((Sun-O)))* (Waitākere Ranges, 2022), is a limited edition handsewn book now housed at the Bodleian Library, University of Oxford.

TIM SAUNDERS farms sheep and beef in the Manawatū. He has had poetry and short stories published in journals in New Zealand and around the world and was shortlisted for the 2021 Commonwealth Short Story Prize. He has written two books: *This Farming Life* (Allen & Unwin, 2020) and *Under a Big Sky* (Allen & Unwin, 2022).

ABOUT THE POETS

KEITH SINCLAIR (1922–1993) was a poet, biographer and historian. His *A History of New Zealand* (1957) profoundly influenced mid-twentieth-century conceptions of nationhood. He was the driving force behind the 1967 establishment of the *New Zealand Journal of History,* remaining its editor until 1987, and was influential in the development of the *Dictionary of New Zealand Biography.* He published five collections of poetry, many with historical themes.

ELIZABETH SMITHER has published 19 collections of poetry as well as novels, short stories and journals. Her latest collection, *My American Chair*, was published by Auckland University Press in 2022.

RUBY SOLLY (Waitaha, Kāi Tahu, Kāti Māmoe) is a writer, music therapist, taonga pūoro practitioner and doctor of public health living in Pōneke Wellington on the old riwai plantation of her ancestors. She has two books of poetry with Te Herenga Waka University Press: *Tōku Pāpā* and *The Artist*.

J.C. [JACQUELINE CECILIA] STURM (Taranaki, Te Whakatōhea) (1927– 2009) was a poet and short story writer. In 1955, 'For all the Saints', was the first short story written in English by a Māori writer published in *Te Ao Hou*. As literary executor for her husband, the poet James K. Baxter, she channelled all proceeds from his estate into the James K. Baxter Charitable Trust. Much of her prose and poetry portrays a strong sense of alienation from a society fostering inequality and marginalisation of Māori.

JILLIAN SULLIVAN's 13 published books include creative non-fiction, novels and poetry. Her awards include the Juncture Memoir Award in America, the NZSA Beatson Fellowship, and the Kathleen Grattan Poetry Award. She is an earth plasterer and sexual consent activist. Her latest book is *Map for the Heart: Ida Valley essays* (Otago University Press, 2021).

APIRANA TAYLOR (Te Whānau ā Apanui, Ngāti Porou, Ngāti Ruanui) is a nationally and internationally published poet, playwright, short story writer and novelist. He has been the writer in residence at Canterbury and Massey universities, has written and published poetry, plays, short stories, and novels, and has been included in many anthologies. He tours schools, tertiary institutions, universities, marae, galleries and prisons, reading his poetry.

NGAHUIA TE AWEKOTUKU MNZM (Te Arawa, Tūhoe, Ngāpuhi, Waikato) grew up in the steam of Ohinemutu Rotorua. She makes poetry and fiction. She has just published an entertaining yet edifying political/personal memoir, *Hine Toa: A story of bravery*. An accomplished scholarly writer, she is trying to become a poet again.

PĀORA TE PŌTANGAROA (d. 1881) was a prophet and rangatira of the Rangitāne iwi in the Wairarapa.

TE KOOTI ARIKIRANGI TE TŪRUKI (c. 1820–1891) was one of the foremost Māori leaders of the nineteenth century. He was a Rongowhakaata chief, military leader, prophet and religious founder. He fought a war against land confiscation and illegal land purchases. In 1868, he founded the Ringatū church, which continues to this day.

ANTHONIE TONNON is a songwriter and performer originally from Ōtepoti Dunedin and now based in Whanganui. His album *Leave Love Out Of This* won the 2022 Taite Music Prize. He was the New Zealand Geographical Society's 2022 Honorary Geographer for his song writing and his public transport advocacy in Aotearoa New Zealand.

DENYS TRUSSELL was born in Ōtautahi Christchurch in 1946. In 1999, his fifth volume of poetry *Walking into the Millennium and Shorter Poems* was short-listed in the Montana New Zealand Book Awards. His long poem 'Archipelago', was a focus document at the 2000 Symposium on New

Literatures in English in Aachen, Germany. *By Sea Mouths Speaking: Collected poems 1973–2018 and related prose* was published in 2019 and his prose and poetry volume in association with the artist Nigel Brown, *Albatross Neck*, appeared in 2022.

TŪMATAHINA (c. 1475) was a Muriwhenua chief at Murimotu, an island off North Cape. He was killed by Ngāpuhi invaders after saving his people by evacuating them to a cave accessible only by a rope used to cross from the island to the shore. He chose the kuaka or godwit to personify his people in the whakataukī or tauparapara 'Ruia, ruia, tahia, tahia'.

BRIAN TURNER ONZM is a poet, essayist, anthologist, journalist, sportsman and ardent conservationist. He was born in Ōtepoti Dunedin and lives in Oturehua in Central Otago. The region is central to his writing, which focuses on the South Island back country landscape and its role in senses of self and belonging. He has published 13 collections of poetry. He was New Zealand Poet Laureate 2003–2005 and in 2020 was appointed an Officer of the New Zealand Order of Merit for services to literature and poetry.

HONE TUWHARE (Ngāpuhi, Te Uri o Hau) (1922–2008) was a boilermaker, poet, short story writer and playwright. He was born at Kokewai, Northland. After losing his mother when he was five, he lived a nomadic life with his father in Kaikohe and Tāmaki Makaurau Auckland, which fostered a strong connection to the working class and social justice. His debut poetry collection, *No Ordinary Sun* (1964), was the first by a Māori poet in English and sold out in 10 days. His 13 subsequent collections were widely read and popularised Māori concepts in English poetry. He was New Zealand Poet Laureate from 1999 to 2001.

TIM UPPERTON is the author of *A House On Fire* (Steele Roberts, 2009), *The Night We Ate The Baby* (Haunui Press, 2016), and *A Riderless Horse* (Auckland University Press, 2022). He won the Caselberg International Poetry Competition in 2012, 2013 and 2020. He lives in Te Papaioea Palmerston North.

MURU WALTERS (Te Aupouri, Te Rarawa) is an Anglican Bishop, master carver, author, poet, composer, artist, broadcaster and former Māori All Black. In 1992, he was ordained Bishop of Upoko o Te Ika. From 1983–2003, he wrote for the Kupu Korikori New Zealand Radio Broadcast. He is highly regarded for his haka compositions.

IAN WEDDE ONZM was born in Te Waiharakeke Blenheim and has lived in a number of different places, including Port Chalmers near Ōtepoti Dunedin; hence, Aramoana, the location of *Pathway to the Sea*. The artist Ralph Hotere was a neighbour and active in protests against the proposed aluminium smelter at Aramoana; he made drawings for the publication of the poem.

ALBERT WENDT ONZ CNZM is a poet, novelist, short story writer, playwright, artist, academic and educator. He was born in Apia, Sāmoa and educated in New Zealand. He was professor of New Zealand literature at Auckland University, the first Pacific person to be appointed an English professor there. He edited several anthologies of Pacific literature. His writing synthesises Polynesian history, myths and oral traditions with contemporary written fiction, and portrays the traditions of the papālagi and their effects on Samoan culture. In 2013, he was appointed a Member of the Order of New Zealand for his pivotal role in the formation of Pacific literature in English.

DORA WILCOX (1873–1953) was a poet and playwright. She was born in Ōtautahi Christchurch and attended Christchurch University College. She was a teacher in Australia and travelled to England where she lived before returning to Australia.

GEORGE PHIPPS WILLIAMS (1846–1909) was an engineer and poet. He was born in London and immigrated to Ōtautahi Christchurch in 1869.

ABOUT THE POETS

ANNE GLENNY WILSON (1848–1930) was a poet and romantic fiction writer. She was born in Victoria, Australia. In 1874 she married James Glenny Wilson, an Australian of Scottish descent, and lived on their 6210 acre farm in the Rangitikei Block near Bulls, known as Ngaio Station.

BRIAR WOOD (Ngāpuhi) is an internationally published writer whose writing resonates with ecological concerns. Her collection of poems *Rāwāhi* (Anahera Press, 2017) was shortlisted for the 2018 Ockham New Zealand Book Awards Poetry Award. *A Book of Rongo and Te Rangahau* (Anahera Press) was published in 2022.

SUE WOOTTON lives in Ōtepoti Dunedin. Her fifth and most recent collection of poetry, *The Yield* (Otago University Press, 2017), was a finalist for the Mary and Peter Biggs Award for Poetry in the 2018 Ockham New Zealand Book Awards. Her poetry has been recognised in a range of other literary awards including the New Zealand Poetry Society International Poetry Competition, the International Hippocrates Prize for Poetry and Medicine, the Gwen Harwood Poetry Award and the University of Canberra Vice-Chancellor's International Poetry Prize. She was the 2008 Robert Burns Fellow at the University of Otago, and in 2023 she travelled to France as the 50th New Zealand writer to hold the Katherine Mansfield Menton Fellowship.

Sources

Acknowledgements for permission to reprint items included in this anthology are made to the following copyright holders and places of publication. Every effort has been made to find the correct copyright holders and estates, but if there are omissions or oversights, we apologise, and in these cases we would welcome contact from the poets or their families.

The early years

Bathgate, Alexander, 'To the Makomako, or Bell-Bird (Now rapidly dying out of our land)', in *A Treasury of New Zealand Verse*, eds W.F. Alexander and A.E. Currie (Auckland: Whitcombe and Tombs, 1926)

Baughan, Blanche, 'A Bush Section', in *Shingle Short and Other Verses* (Christchurch: Whitcombe and Tombs, 1908)

Bethell, Ursula, 'Pause', in *From a Garden in the Antipodes* by Evelyn Hayes (London: Sidgwick and Jackson, 1929)

Hapai, Horomona, 'He Tangi mo te Matenga o Ngā Kai (Lament for a Failed Crop)', in *Māori Poetry: An introductory anthology*, ed. Margaret Orbell (Auckland: Heinemann Educational, 1978)

Hutchinson, Francis, 'Drought', in *New Zealand Farm and Station Verse, 1850–1950*, ed. A.E. Woodhouse (Christchurch: Whitcombe and Tombs, 1950)

Mulgan, Alan E., 'Dead Timber', in *A Treasury of New Zealand Verse*, eds W.F. Alexander and A.E. Currie (Auckland: Whitcombe and Tombs, 1926)

Pua-Roro, 'He Waiata Whakaaraara Pā (A Sentinel's Song)', in *Ngā Mōteatea*, ed. Ā.T. Ngata (Wellington: Polynesian Society, 1944–57)

Reeves, William Pember, 'The Passing of the Forest', in *A Treasury of New Zealand Verse*, eds W.F. Alexander and A.E. Currie (Auckland: Whitcombe and Tombs, 1926)

Reeves, William Pember and George Phipps Williams, 'An Old Chum on New Zealand Scenery', in *Colonial Couplets: Being Poems in Partnership*, eds George Phipps Williams and W.P. Reeves (Christchurch: Simpson & Williams, 1889)

Wilcox, Dora, 'In London', in *A Treasury of New Zealand Verse*, eds W.F. Alexander and A.E. Currie (Auckland: Whitcombe and Tombs, 1926)

Wilson, Anne Glenny, 'The Forty-Mile Bush', in *A Treasury of New Zealand Verse*, eds W.F. Alexander and A.E. Currie (Auckland: Whitcombe and Tombs, 1926)

The middle years

Adcock, Fleur, 'Last Song', in *The Incident Book* (Oxford and New York: Oxford University Press, 1986)

Baxter, James K., 'Waipatiki Beach', in *Pig Island Letters* (London: Oxford University Press, 1966)

Bland, Peter, 'Beginnings', in *Primitives: Poems by Peter Bland* (Wellington: Wai-te-ata Press, 1979)

Blank, Arapera Hineira Kaa, 'Conversation with a Ghost 1974–1985 (on looking at other people's houses)', in *Nga Kokako Huataratara: The notched plumes of the Kokako* (Auckland: Waiata Koa Trust, 1986)

Brasch, Charles, 'The Land and the People (III)', in *The Land and the People and Other Poems* (Christchurch: Caxton Press, 1939)

Campbell, Alistair Te Ariki, 'The Return', in *Mine Eyes Dazzle: Poems 1947–49* (Christchurch: Pegasus Press, 1950)

Carter, Jacqueline, 'Our Tūpuna Remain', in *Puna Wai Kōrero: An anthology of Māori poetry in English*, eds Reina Whaitiri and Robert Sullivan (Auckland: Auckland University Press, 2014)

Curnow, Allen, 'House and Land', in *The Penguin Book of New Zealand Verse*, ed. Allen Curnow (Auckland: Penguin, 1960) [Permission is courtesy of the copyright owner, Tim Curnow, Sydney, care of Auckland University Press]

Dallas, Ruth, 'Pioneer Woman With Ferrets', in *Walking on the Snow* (Christchurch: Caxton Press, 1976)

Dansey, Harry, 'The Old Place', in *Puna Wai Kōrero: An anthology of Māori poetry in English,* eds Reina Whaitiri and Robert Sullivan (Auckland: Auckland University Press, 2014) [Permission is courtesy of the Dansey family]

Edmond, Lauris, 'Atom Bomb Test, Moruroa Atoll, 6 September 1995', in *Below the Surface*, ed. Ambury Hall (Auckland: Vintage NZ, 1995)

Glensor, Patricia, 'Otago Landscape', in *Private Gardens: An anthology of New Zealand women poets,* ed. Riemke Ensing (Dunedin: Caveman Press, 1986)

Glover, Denis, 'Lake Manapouri', in *Dancing to my Tune* (Wellington: Catspaw Press, 1974) [Permission is courtesy of the Denis Glover Estate and Pia Glover, the copyright holder]

Harrison, Rangi T., 'Waikato te Awa (Waikato is the River)', in 'He kōrero paki nō tawhiti mai: Narratives from distant past te wai ka tō hia he wai mā ū; Caught by the drag of the water', by Herewini Easton (Master's thesis, Auckland University of Technology, 2020)

Hawken, Dinah, 'Hope', in *Water, Leaves, Stones* (Wellington: Victoria University Press, 1995)

Hood, Noleen, 'Blaketown Beach', in *Trim Pork and Waterfalls* (Blaketown: Ledingham Books, 1991)

Hooper, Peter, 'Three Pines by the Hohonu', in *Fragments III: Earth Marriage* (Christchurch: David Young & David Waddington, 1972) [Permission courtesy of Brian Turner, executor]

Hulme, Keri, 'Te Rapa, Te Tuhi, Me Te Uira (or Playing with Fire)', in *Below the Surface*, ed. Ambury Hall (Auckland: Vintage NZ, 1995)

Hunt, Sam, 'A Bottle Creek Blues', in *From Bottle Creek* (Wellington: Alister Taylor, 1972)

McDonald, Donald, 'Beginning Again', in *New Zealand Farm and Station Verse 1850–1950*, ed. A.E. Woodhouse (Christchurch: Whitcombe and Tombs, 1950)

McQueen, Cilla, 'The Mess We Made at Port Chalmers', in *Crikey* (Dunedin: John McIndoe, 1994)

Melbourne, Hirini, 'Aramoana', in 'He kupu tuku iho mō tēnei reanga: A critical analysis of waiata and haka as commentaries and archives of Māori political history', by Rachael Te Āwhina Ka'ai-Mahuta (PhD thesis, Auckland University of Technology, 2010)

Menzies, Trixie Te Arama, 'Papakainga', in *Papakainga* (Auckland: Waiata Koe, 1988)

Mitcalfe, Barry, 'The Road', in *Look to the Land* (Coromandel: Coromandel Press, 1988)

Murray, Saana, 'Te Kōkota o Pārengarenga (The White Sands of Pārengarenga)', in *Toiapiapi – He huinga o ngā kura puoro a te Māori: A collection of Māori musical treasures* by Hirini Melbourne (Wellington: Shearwaters, 1993)

Newton, John, 'Eldorado Poem', in *Tales from the Angler's Eldorado* (Christchurch: Untold Books, 1985)

Orsman, Chris, 'Ornamental Gorse', in *Ornamental Gorse* (Wellington: Victoria University Press, 1994)

Patuawa-Nathan, Evelyn, 'Waikato Lament', in *Opening Doors* (Suva: Mana Publications, 1979)

Penfold, Merimeri, 'Tamaki-makau-rau / Tamaki of a Hundred Lovers', in *The Penguin Book of New Zealand Verse*, eds Ian Wedde and Harvey McQueen (Auckland: Penguin, 1985)

Pere, Vernice Wineera, 'Song from Kapiti', in *Mahanga: Pacific Poems* (Hawai`i: The Institute for Polynesian Studies, Brigham Young University, 1978)

Potiki, Roma, 'A Cloak and Taiaha Journey', in *Shaking the Tree* (Wellington: Steele Roberts, 1998)

Ricketts, Harry, 'Memo for Horace', in *Below the Surface*, ed. Ambury Hall (Auckland: Vintage NZ, 1995)

Sinclair, Keith, 'The Bomb is Made', in *A Time to Embrace* (Auckland: Paul's Book Arcade, 1963)

Sturm, J.C., 'As the Godwits Fly', in *Dedications* (Wellington: Steele Roberts, 1996)
Taylor, Apirana, 'Feelings and Memories of a Kuia', in *Te Ao Mārama: Contemporary Māori writing* vol. 1, ed. Witi Ihimaera (Auckland: Reed Books, 1992)
Te Awekotuku, Ngahuia, 'Mururoa/Moruroa', in *Below the Surface*, ed. Ambury Hall (Auckland: Vintage NZ, 1995)
Trussell, Denys, 'Ore: An ecological poem', in *Words for the Rock Antipodes* (Auckland: Hudson Cresset, 1986)
Turner, Brian, 'Van Morrison in Central Otago', in *Beyond* (Dunedin: John McIndoe, 1992)
Tuwhare, Hone, 'No Ordinary Sun', in *No Ordinary Sun* (Wellington: Blackwood and Janet Paul, 1964)
—, 'Papa-tu-a-Nuku (Earth Mother)', in *Making a Fist of It: Poems & short stories* (Dunedin: Jackstraw Press, 1978)
Walters, Muru, 'Haka: He huruhuru toroa / Haka: The feathered albatross', in *The Penguin Book of New Zealand Verse*, eds Ian Wedde and Harvey McQueen (Auckland: Penguin, 1985)
Wedde, Ian, 'Pathway to the Sea', in *Pathway to the Sea* (Christchurch: Hawk Press, 1975)
Wendt, Albert, 'In the Midnight Ocean of His Sleep', in *Shaman of Visions* (Auckland: Auckland University Press and Oxford University Press, 1984)

Now

Argante, Jenny, 'Seal Mourning', in *Our Own Kind: 100 New Zealand poems about animals*, ed. Siobhan Harvey (Auckland: Godwit, 2009)
Auchmuty, Bridget, 'Marginalia', in *Unmooring* (Christchurch: Quentin Wilson, 2020)
Avia, Tusiata, 'Fucking St Barbara (i)', in *The Savage Coloniser Book* (Wellington: Victoria University Press, 2020)
Baker, Hinemoana, 'Huia, 1950s', in *mātuhi / needle* (Wellington: Victoria University Press, 2004)
Beautrais, Airini, 'Trout / *Oncorhynchus mykiss* / *Salmo trutta*', in *Flow: Whanganui River poems* (Wellington: Victoria University Press, 2017)
Burke, Brenda, 'Feature Battle', in *Kaupapa: New Zealand poets, world issues*, eds Hinemoana Baker and Maria McMillan (Wellington: Development Resource Centre, 2007)
Carter, Jacq, 'If I Am the River and the River is Me', in *Puna Wai Kōrero: An anthology of Māori poetry in English*, eds Reina Whaitiri and Robert Sullivan (Auckland: Auckland University Press, 2014)
Cweorth, Jonathan, 'Ghost Stoat', in *Manifesto Aotearoa: 101 political poems*, eds Philip Temple and Emma Neale (Dunedin: Otago University Press, 2017)
Davidson, Lynn, 'Speaking to the Otter', in *Islander* (Wellington and Oxfordshire: Victoria University Press and Shearsman Books, 2019)
Eggleton, David, 'A Report on the Ocean', in *The Spinoff: Friday Poem*, 28 January 2022, thespinoff.co.nz
Faith, Rangi, 'Losing our Mana', in *Conversation with a Moa Hunter* (Wellington: Steele Roberts, 2005)
Farrell, Fiona, 'Eel', in *Turbine / Kapohau 05* (Wellington: Victoria University Press, 2005)
Fitchett, Sue, 'Story Lines', in *Manifesto Aotearoa: 101 political poems*, eds Philip Temple and Emma Neale (Dunedin: Otago University Press, 2017)
Glenny, Alison, 'Candle', in *Bird Collector* (Auckland: Compound Press, 2021)
Hamel, Jordan, 'Te Aro', in *Glass: A journal of poetry* (Toledo, OH: Glass Poetry Press, June 2019)
Hawken, Dinah, 'Losing Everything', in *There is No Harbour* (Wellington: Victoria University Press, 2019)
Hawkes, Rebecca, 'The Land without Teeth', in *AUP New Poets 5* (Auckland: Auckland University Press, 2019)
Heath, Helen, 'The Anthropocene *circa 2016*', in *Are Friends Electric?* (Wellington: Victoria University Press, 2018)
Howell, John, 'Old Bones', in *Manifesto Aotearoa: 101 political poems* (Dunedin: Otago University Press, 2017)
Ingram, Gail, 'Recipe for a Unitary State', in *Manifesto Aotearoa: 101 political poems* (Dunedin: Otago University Press, 2017)

Ireland, Kevin, 'Moruroa: The name of the place', in *Fourteen Reasons for Writing: New poems* (Christchurch: Hazard Press, 2001)
Jackson, Anna, 'Huia', in *The Pastoral Kitchen* (Auckland: Auckland University Press, 2002)
Jane, Ash Davida, '2050', in *How to Live with Mammals* (Wellington: Victoria University Press, 2021)
Jones, Tim, 'All that Summer', in *New Sea Land* (Wellington: Mākaro Press, 2016)
Kennedy, Anne, 'Flood Monologue', in *The Sea Walks into a Wall* (Auckland: Auckland University Press, 2021)
Kennedy, Erik, 'Phosphate from Western Sahara', in *Another Beautiful Day Indoors* (Wellington: Te Herenga Waka University Press, 2022)
Kitching, Megan, 'I. Pūhā', in *At the Point of Seeing* (Dunedin: Otago University Press, 2023)
Kyle, Leicester, '*An Aside:* Advice to the Rehabilitator', from 'Death of a Landscape', in *The Millerton Sequences* (Auckland: Atuanui Press, 2014)
Latham, Arihia, 'Birdspeak', in *Birdspeak* (Whanganui: Anahera Press, 2023)
Lehndorf, Helen, 'Oh Dirty River', in *The Comforter* (Wellington: Seraph Press, 2011)
Maihi, Toi Te Rito, 'Korari / Harakeke', in *Whakaaro Aroha* (Northland: Arts Promotion Trust, 2003)
Manhire, Bill, 'Huia', in *Wow* (Wellington: Victoria University Press, 2020)
Marsh, Selina Tusitala, 'Girl from Tuvalu', in *Dark Sparring* (Auckland: Auckland University Press, 2013)
McCurdie, Carolyn, 'Ends', in *Manifesto Aotearoa: 101 political poems* (Dunedin: Otago University Press, 2017)
McMillan, Maria, 'How They Came to Privatise the Night', in *Tree Space* (Wellington: Victoria University Press, 2014)
McQueen, Cilla, 'Frogs', in *Poeta: Selected & new poems* (Dunedin: Otago University Press, 2018)
Mei, Van, 'There's Real Mānuka Honey in Heaven', in *Starling*, summer 2019, starlingmag.com
Mila, Karlo, 'Matariki: A call to kāinga', in *Goddess Muscle* (Wellington: Huia, 2020)
Molloy, Harvey, 'Dear ET', in *Udon by the Remarkables* (Wellington: Makaro Press, 2016)
Morton, Elizabeth, 'Birdlife in a Broken Century', in *Naming the Beasts* (Dunedin: Otago University Press, 2022)
Neale, Emma, 'Huia', in *Tender Machines* (Dunedin: Otago University Press, 2015)
O'Connor, Naomi, 'So You Don't Belong, Pohoot', in *The Nature of Things: Poems from the New Zealand landscape*, ed. James Brown (Nelson: Craig Potton, 2005)
Piahana-Wong, Kiri, 'Piha', in *Tidelines* (Auckland: Anahera Press, 2024)
Pickens, Robyn Maree, 'Praise the Warming World (Try To)', in *Tung* (Dunedin: Otago University Press, 2023)
Powles, Nina Mingya, 'Whale Fall', in *Sweet Mammalian*, issue 2, 2015.
Rapatahana, Vaughan, 'Topside, Nauru', in *Cordite Poetry Review*, 1 February 2013, cordite.org.au
Reeve, Richard, 'The Old Breed', in *Generation Kitchen* (Dunedin: Otago University Press, 2015)
Rolleston, Te Kahu, 'The Rena', in *Puna Wai Kōrero: An anthology of Māori poetry in English*, eds Reina Whaitiri and Robert Sullivan (Auckland: Auckland University Press, 2014)
Sampson, Sam, 'Erasure', in *Halcyon Ghosts* (Auckland: Auckland University Press, 2014)
Saunders, Tim, 'Dad's Piece of Sky', in *Poetry Notes*, vol. 11, issue 3 (Wellington: Poetry Archive of New Zealand Aotearoa, Summer 2022)
Smither, Elizabeth, 'Port Hills, Canterbury', in *Landfall* 242 (Dunedin: Otago University Press, November 2021)
Solly, Ruby, 'River Songs – Waimāpihi', in *Tōku Pāpā* (Wellington: Victoria University Press, 2021)
Sullivan, Jill, 'Choosing', in *Parallel* (Wellington: Steele Roberts, 2014)
Taylor, Apirana, 'my whenua', in *Te Ata Kura: The red-tipped dawn* (Christchurch: Canterbury University Press, 2004)
Tonnon, Anthonie, 'Mataura Paper Mill', in *Leave Love Out of This*, 2021, anthonietonnon.bandcamp.com
Turner, Brian, 'Lament for the Taieri River', in *Into the Wider World: A back country miscellany* (Auckland: Random House NZ, 2008)

Tuwhare, Hone, 'Purangi (Kauri Snail shell)', in *Small Holes in the Silence: Collected poems* (Auckland: Random House NZ, 2016)

Upperton, Tim, 'Manawatū', in *A Riderless Horse* (Auckland: Auckland University Press, 2022)

Wood, Briar, 'Whai', in *Rāwāhi* (Auckland: Anahera Press, 2017)

Wootton, Sue, 'A Behoovement', in *The Yield* (Dunedin: Otago University Press, 2017)

Further reading

Bethell, Ursula, *Collected Poems* (Christchurch: Caxton Press, 1950)

Campbell, Alistair Te Ariki, *The Collected Poems of Alistair Te Ariki Campbell* (Wellington: Victoria University Press, 2016)

Glenny, Alison, *The Farewell Tourist* (Dunedin: Otago University Press, 2018)

Glover, Denis, *Clutha: River poems* (Dunedin: John McIndoe Press, 1977)

—, *Enter Without Knocking: Selected poems* (Christchurch: Pegasus Press, 1964)

Hawken, Dinah, *It Has No Sound and is Blue* (Wellington: Victoria University Press, 1987)

—, *Ocean and Stone* (Wellington: Victoria University Press, 2015)

—, *Oh THERE you are TUI! New and selected poems* (Wellington: Victoria University Press, 2001)

Hulme, Keri, *The Silences Between (Moeraki Conversations)* (Auckland: Auckland University Press and Oxford University Press, 1982)

McNaughton, Trudie, ed., *Countless Signs: The New Zealand landscape in literature: An anthology* (Auckland: Reed Methuen, 1986)

Mila, Karlo, *Dream Fish Floating* (Wellington: Huia, 2005)

Newman, Janet, 'Embodiment and Solace: The entanglement of culture with nature in contemporary Aotearoa New Zealand Ecopoetry', in *Poetry and the Global Climate Crisis: Creative educational approaches to complex challenges*, eds Amatoritsero Ede et al. (Oxfordshire: Routledge, 2024), pp. 10–26

—, 'Thinking like a leaf: Dinah Hawken, Romantic ecopoet', *Journal of New Zealand Literature* 35, 1, 2017, pp. 8–27

Piahana-Wong, Kiri, *Night Swimming* (Auckland: Anahera Press, 2013)

Potiki, Roma, *Oriori: A Māori child is born – from conception to birth* (Auckland, Tandem Press, 1999)

—, *Stones in Her Mouth* (Auckland: Ihimaera Williams Associates, 1992)

Reeve, Richard, *Dialectic of Mud* (Auckland: Auckland University Press, 2001)

—, *In Continents* (Auckland: Auckland University Press, 2008)

—, *The Life and the Dark* (Auckland: Auckland University Press, 2004)

Ross, Jack, ed. *Poetry New Zealand Yearbook 2019* (Palmerston North: Massey University Press, 2019)

Sullivan, Robert, *Shout Ha! To the Sky* (Bloomsbury: Salt Publishing, 2010)

—, *Star Waka* (Auckland: Auckland University Press, 1999)

—, *Tūnui/Comet* (Auckland: Auckland University Press, 2022)

—, *Voice Carried My Family* (Auckland: Auckland University Press, 2005)

— et al., *Puna Wai Korero: An anthology of Maori poetry in English* (Auckland: Auckland University Press, 2014)

— et al., *Whetu Moana: Contemporary Polynesian poems in English* (Auckland: Auckland University Press, 2002)

Taylor, Apirana, *Eyes of the Ruru* (Wellington: Voice Press, 1979)

—, *Soft Leaf Falls of the Moon* (Auckland: Pohutukawa Press, 1999)

Trussell, Denys, *By Sea Mouth Speaking: Collected poems 1973–2018 and related prose* (Auckland and Palm Springs: Brick Row, 2019)

Tuwhare, Hone, *Come Rain Hail* (Dunedin: Otago University Press, 1970)

—, *Making a Fist of It: Poems & short stories* (Dunedin, Jackstraw Press, 1978)

Turner, Brian, *Bones* (Dunedin: John McIndoe, 1985)

—, *Ladders of Rain* (Dunedin: John McIndoe, 1978)

Acknowledgements

My mum planted a kauri tree in our driveway and showed me that with care, even northern growers could be coached into life beyond 38 degrees south. This year it grew its first cones. My dad taught me to tend cattle, to tread lightly. Poets taught me there are many ways to walk the earth, many ways to think it. I am grateful to them all. The most important acknowledgements must go to the poets and copyright holders who allowed work to be published in this anthology. Many of the contributors are established poets who have published widely, and it is a privilege to be allowed to include their work here. I offer sincere thanks to all the poets who generously permitted me to reproduce their work in this book. Grateful acknowledgments are due to the families and copyright holders of late writers for allowing the use of their relative's work.

I'm grateful to Massey University Professor Bryan Walpert for suggesting the idea of an Aotearoa New Zealand ecopoetry anthology in September 2016 at the Aotearoa Creative Writing Research Network Colloquium, Ahi Ka: Building the Fire. At that time, I was curating this country's ecopoetry according to a baseline of generally accepted and soon-to-be-disrupted ecopoetics. Bryan said an anthology would be easy, by which I think he meant doable. He lit the spark.

I'm grateful to former Massey University Associate Professor Ingrid Horrocks, who suggested I look into the relatively new field of postcolonial ecocriticism as part of my doctoral research in 2018. By then, I thought my thesis was nearly finished, yet here was the ground in which it is rooted and which underlies the context for Aotearoa New Zealand ecopoetry as presented in this anthology.

I am grateful to Otago University Press publisher Sue Wootton, who believed in the anthology from the beginning and, with wisdom and patience, helped me turn the idea into a book.

My deepest gratitude to poet and academic Dr Robert Sullivan, whose poetry showed me connections between ecologies and people across time and place, which expanded my conceptions of culture and nature, and who, by agreeing to co-edit this anthology, gave me support, confidence, knowledge and tikanga.

This anthology would not have been doable without poet and editor Dr Margaret Moores, who, when I said the job was too big, offered to help. Margaret spent 2022 researching from her home and, with steadfastness and spreadsheets, made the task seem smaller. I am immensely grateful for her assiduous research and close readings of the text.

My grateful thanks to Tara Jahn Werner for the use of her comprehensive poetry library.

Thanks to Otago University Press editor Mel Stevens, who sought permissions for the poems and digitised those that previously existed only on paper, and to Otago University Press production manager Fiona Moffat for outstanding design and production.

Special thanks to artist Bing Dawe for permission to use on the anthology's cover his impactful artwork showing nature's resilience, persistence and need of attention.

Love and thanks to my family, Frank Taylor, Lyle, Wynton and Delia Newman, my sister Sylvia and the Clutterbucks, and my parents, Doug and Ethel Newman, who, even in memory, keep me grounded.

Janet Newman

First I wish to thank Janet Newman for her assiduous research, vision and commitment to the cause of an ecopoetry anthology reflecting diverse voices from Aotearoa New Zealand. This anthology would not exist without her extensive work here. Ngā mihi nui ki a koe. I wish to acknowledge my whānau from Ngāti Manu in Kāretu and my whānau from Puketeraki Marae and the Huriawa Pā in East Otago. Thanks to my mother, Maryann, my tamariki Temuera, Eileen and Turi, and the wider Sullivan and Conlon whānau. Finally, my thanks to the poets whose work is represented here, and those whose work is not included but on whose shoulders we stand for seeing the world in all its wondrous complexity, and who confront the incomprehensible loss of our healthy world. A historian might describe this devastating loss as arriving through colonisation, the military industrial complex, and transnational neocapitalism. A poet might reach into the soil, look into the ocean, give hands up to the sky, and speak with the creatures there. Tēnei te mihi kau ana ki a koutou katoa e ngā kaituhi toikupu.

Arohanui.

Robert Sullivan

Published by Otago University Press
Te Whare Tā o Ōtākou Whakaihu Waka
533 Castle Street
Dunedin, New Zealand
university.press@otago.ac.nz
oup.nz

First published 2024
Volume copyright © Otago University Press
Text copyright © Individual poets as listed on contents page
The moral rights of the editors have been asserted

ISBN 978-1-99-004881-4

Published with the assistance of Creative New Zealand

A catalogue record for this book is available from the National Library of New Zealand. This book is copyright. Except for the purpose of fair review, no part may be stored or transmitted in any form or by any means, electronic or mechanical, including recording or storage in any information retrieval system, without permission in writing from the publishers. No reproduction may be made, whether by photocopying or by any other means, unless a licence has been obtained from the publisher.

Editor: Mel Stevens
Cover: Bing Dawe, *Head of a shag*, from the series 'Head count', 2007. Acrylic on paper, 49.5 x 35 cm.

Printed in New Zealand by Caxton